IMPRINT IN HUMANITIES

The humanities endowment
by Sharon Hanley Simpson and
Barclay Simpson honors
MURIEL CARTER HANLEY
whose intellect and sensitivity
have enriched the many lives
that she has touched.

The publisher and the University of California Press Foundation gratefully acknowledge the generous support of the Simpson Imprint in Humanities.

Transforming Psychological Worldviews to Confront Climate Change

Transforming Psychological Worldviews to Confront Climate Change

A Clearer Vision, A Different Path

F. STEPHAN MAYER

UNIVERSITY OF CALIFORNIA PRESS

University of California Press, one of the most distinguished university presses in the United States, enriches lives around the world by advancing scholarship in the humanities, social sciences, and natural sciences. Its activities are supported by the UC Press Foundation and by philanthropic contributions from individuals and institutions. For more information, visit www.ucpress.edu.

University of California Press
Oakland, California

Library of Congress Cataloging-in-Publication Data

Names: Mayer, F. Stephan, author.
Title: Transforming psychological worldviews to confront climate change : a clearer vision, a different path / F. Stephan Mayer.
Description: Oakland, California : University of California Press, [2019] | Includes bibliographical references and index. |
Identifiers: LCCN 2018024586 (print) | LCCN 2018026793 (ebook) | ISBN 9780520970571 (e-edition) | ISBN 9780520298460 (cloth) | ISBN 9780520298453 (pbk.)
Subjects: LCSH: Climatic changes—United States—Psychological aspects. | Environmental ethics.
Classification: LCC BF353.5.C55 (ebook) | LCC BF353.5.C55 M37 2019 (print) | DDC 155.9/16—dc23
LC record available at https://lccn.loc.gov/2018024586

27 26 25 24 23 22 21 20 19 18
10 9 8 7 6 5 4 3 2 1

To Kristen, Caleb, Rachael, Eric, and Ruby

BRIEF CONTENTS

Foreword *xv*

Prologue *xix*

1. See Better: Psychology as a Foundational Science to Confront Climate Change *1*
2. The Fundamental Problem: The Psychology behind Climate Change *29*
3. The Emergency of Climate Change: Why Are We Failing to Take Action? *67*
4. The Great Transition: From Separateness to Interconnectedness *108*
5. Actions Being Taken to Transition to the Land Ethic Worldview *151*

References *175*

Index *191*

DETAILED CONTENTS

FOREWORD xv

PROLOGUE xix

1. SEE BETTER: PSYCHOLOGY AS A FOUNDATIONAL SCIENCE TO CONFRONT CLIMATE CHANGE

Averting a Present-Day Tragedy 2

Envisioning Psychology as a Foundational Science to Confront the Environmental Threat of Climate Change 9

The Psychology of Perception versus Reality 12

Schemas and Misperceptions 15

Cultural Worldviews, Cultural Schemas, and Unjustified Clarity Distortions 17

Do Cultural Schemas Help or Hinder Us? 19

The Issue of Critiquing Worldviews 21

The Emphasis within the Environmental Movement: Is it Severely Limited? 24

In Summary: Five Major Issues 25

Final Thoughts 26

2. THE FUNDAMENTAL PROBLEM: THE PSYCHOLOGY BEHIND CLIMATE CHANGE

Unjustified Clarity Distortions: A Further Elaboration 29

Why Critique the United States' Worldview? 33

The General Argument 34
Feelings of Separation, Distance, and Indifference 34
Cultural Striving for Superiority and Self-Enhancement 37
Articulating and Critiquing the Worldview of the United States 40
The Individualistic Cultural Lens: The Autonomous Self and Distancing 40
Feelings of Superiority: Hierarchy and Distancing 44
The Big Picture: Shifting from an Ego to an Eco Orientation 52
The Idea of Progress: Striving for Personal Success and Material Wealth 55
Analytic as Opposed to Holistic Thinking: Seeing Parts, Not Wholes 61
The Belief in a Just World: Moral Distancing and Denial of Impropriety 62
Final Thoughts 64
Seeing Clearly Means Seeing Connections 65
Looking Ahead 66

3. THE EMERGENCY OF CLIMATE CHANGE: WHY ARE WE FAILING TO TAKE ACTION?

Introducing Latane and Darley's Model of Prosocial Behavior 68
The Psychology behind Proenvironmental Behavior 72
Stage 1. Noticing the Event 75
Sensory Limitations 75
Our Ancient Brain 76
Urban Lifestyle / Living Indoors 77
Technological Devices 78
Being Self-Absorbed 78
Lack of Place Attachment 79
Habituation 80
Escapist Activities 80
General Ignorance 80
Stage 2. Interpreting the Event as an Emergency 81
Incongruity of Pleasant Weather as Threat 81
Optimism Bias 82
Technosalvation / Suprahuman Powers 83
Discounting the Messenger / Mistrust 84
Right-Wing Political Ideology 85

Disinformation Campaigns 87
A General Atmosphere of Uncertainty 88
Social Comparison 88
Selective Exposure 89
Social Norms / Pluralistic Ignorance 90
Denial 91
Psychological Distance 92

Stage 3. Feeling Personal Responsibility and a Sense of "We" 92
Mastery-Oriented Individualistic Worldview 93
Lack of Place Attachment 94
Factors Influencing Causal Attributions 94
Diffusion of Responsibility 95
Temporal/Spatial Proximity 95
Self-Enhancing Tendencies 95
Just-World Beliefs 96
Ethnocentric Beliefs 96
Magnitude of Cause and Effect 96

Stages 4 and 5. Forming an Idea of What to Do and Having the Ability to Do It 97
Feeling Overwhelmed 97
Ignorance 98
Fatalism 98
Perceived Self-Efficacy and Collective Efficacy 98
Depleted Directed Attention 99
Lack of Creativity 99
An Emphasis on Cleverness and Analytic Thinking 100
Habit/Commitment 100
Rebound Effect 101
Mistrust 101

The Overriding Cost/Benefit Analysis 101
Risks 101
Financial Investments 103
Perceived Inequity 103
System Justification 103

Conflicting Values, Goals, Aspirations 103
Tokenism 104

Final Thoughts: How Can We Overcome the Obstacles to Helping? 104
Being Realistic about the Opposition and Barriers 104
The Great Transition 106

4. THE GREAT TRANSITION: FROM SEPARATENESS TO INTERCONNECTEDNESS

The Land Ethic Worldview 110
Measuring the Land Ethic Worldview: The Connectedness to Nature Scale 114
Other Related Measures of Connectedness to Nature 115
Final Comments on Scales 122

The Land Ethic, Self-Enhancement, and Self-Transcendence 126

The Land Ethic and Distancing Effects 127

The Issue of Materialism 129

The Land Ethic, Perspective-Taking, and Environmental Concern 130

The Land Ethic and Proenvironmental Behavior 131
Lifestyle/Consumer Practices 132
Support for and Involvement in Environmental Organizations 136
Political Activism on Behalf of the Environment 137

The Land Ethic and Well-Being 138

The Land Ethic and Problem Solving, Holistic Thinking, and Creativity 143

The Land Ethic and Spirituality 144

Final Thoughts 145
Creating Greater Harmony between Human and Environmental Systems 145
Creating Greater Harmony between People 146
The Value of Interconnectedness 147
Reflections on the Great Transition 148
A Word of Caution 149
Can It Be Done? 149

5. ACTIONS BEING TAKEN TO TRANSITION TO THE LAND ETHIC WORLDVIEW

The Oberlin Project 152

Transition Town Totnes 160

Everyday Transitions 164

The Children and Nature Network 165

Urban Planning 167

The Role of Churches 168

Final Thoughts 170

REFERENCES 175

INDEX 191

FOREWORD

The problem is not nature, but human nature and specifically how we think relative to the natural systems on which we depend. The natural world has no need for a pretentious, large-brained, upstart, and destructive keystone species. Measured by the health and resilience of the ecosphere, the earth might do much better with some other species. The cure for what ails us begins with a change in how we think and what we think about. Whether that might require a return to some earlier mode of thought or an evolutionary breakthrough or a combination of the two poses interesting but unanswerable questions. It's not likely that we can return to indigenous ways of thinking in a radically changed, overcrowded, and technologically driven world. It would, however, be prudent to acquire a dose of respect or reverence for nature even if it's just to hedge our bets in light of our overwhelming ignorance. But what would it mean for us to make peace with nature and adjust our expectations and behavior to its limits plus a margin for error?

We cannot begin to think clearly about that complex question unless we first acknowledge that hardly any society deemed to be "primitive" has resisted the seductions of technology—that is, "those things that once possessed cannot be done without," as Wallace Stegner once put it. We have, in effect, a psychological immune deficiency that makes us suckers for new gadgets, whether Winchester rifles or refrigerators or iPhones. Constant technological change, however, means that the challenge of improving our collective prospect for a long tenure on earth is a moving target. But the challenges of fitting ourselves to the ecosphere have been growing ever since we first emerged on the plains of East Africa. Since that brief geological moment, the microscopic problems of bygone ages have grown into large, complex, and earth-shattering crises of the twenty-first century on a crowded and technology-dominated planet. Now it's all or nothing, and the clock, as Professor Mayer notes, is ticking.

If we intend to improve our prospects, the gap between what we know about human psychology and our actual behavior must quickly be narrowed. As this book reports, psychologists have uncovered a great deal about what makes us tick. We are vulnerable to group pressures, perceptual biases, and fears and phobias that lurk in the deep recesses of our ancient reptilian brainstem. We are tribal, prone to demonize outsiders. We hold grudges. We inflict harm for ephemeral reasons. But we also cooperate, care, nurture, write poetry, philosophize, make music, and heal. The difference between these behaviors is attributable in part to the fact that the brain of *Homo sapiens* is a concatenation of conflicting parts that evolved by happenstance to meet the challenges of simpler times. As a result we have evolved into remarkably quick thinkers but slow learners—far more clever than wise.

Beginning with the work of Sigmund Freud, the discipline of psychology has rapidly grown in stature. Given that, and the discipline of neuroscience, it is fair to say that we know a great deal about the mechanics of the brain. I suspect, however, that the mind remains as elusive as ever, and perhaps it always will be a mystery. That's the rub: we know better than we act, and the difference between the two takes us into the murky ground of will and intention, which I will leave to others.

For those, like Professor Mayer, who intend to apply the insights and thoughtful conjectures of psychology to improve the human prospect, the difficulties are many. In an age when we are bombarded with commercial messages—estimated to be five thousand a day—one must reckon with the results of a massive effort to reshape human psychology into forms and behaviors more convenient for commercial exploitation. How much of our behavior is "natural"; how much is warped to the imperatives of consumption? What does the advent of superintelligent machines mean for humans? Where will we fit among our much smarter electronic and increasingly independent progeny? How will crowdedness affect human behavior as we multiply into a population of 10-billion-plus in a mostly urban world? What might manipulation of our own genetic makeup mean for our behavior and thinking? How will we do in a more densely populated, noisier, more polluted, and hotter world? Will those factors impinge on our very sanity?

A great deal is at stake, and it has much to do with how and how well we think and, therefore, how well we act. It's easy to say that we must become smarter about things that matter, understand the patterns and systems that bind us together in time and space, extend our sense of time to a farther horizon, learn the arts and sciences of sustainable living, and become more caring and peaceful. But it is difficult to say how such things can happen at a scale that matters and in the time available. I do think, however, that Professor Mayer is right in saying that the changes in thinking and behavior can begin only in small places, in the homes, neighborhoods, communities, and parts of cities where the scale is manageable and the human relationships are those old and familiar ones revolving around competent parents, solid friendships, cohesive neighbors, and robust civic culture. And here the tools of ecological design wielded by skilled designers, gardeners, foresters, farmers, builders, landscapers, planners, educators, public officials, and engineers can begin the process of re-forming the auto-industrial world into one conducive to conviviality, ecological competence, neighborliness, and beauty that will be the cradle for the next stage of human maturation. If that sounds like utopia, examine carefully some of the evidence cited in this book. It is already happening, and this book is aimed at speeding the process.

David W. Orr

PROLOGUE

This book tackles the environmental threat of climate change from a psychological perspective, an approach that differentiates the book from others on this subject. I do not present different economic or political policies as solutions to this problem; I do not hold up new green technologies as the path to a brighter future. Rather, I argue that our psychological representation of the world, the way we see the world, is at the heart of this issue. The implication of this argument is that we need to change the way we see the world if we are to effectively take a new course of action to address this threat.

On a fundamental level, I argue, climate change is a problem of distorted vision. Namely, on the basis of our worldview we have created human systems (i.e., political and economic systems) that are at odds with environmental systems; the disharmony between these human and environmental systems is the root cause of this threat. The book primarily focuses on the worldview that characterizes the United States. Besides articulating how this view of the world distorts our sight and causes this threat, the book has another main purpose: it also discusses how we might rectify this situation.

All of our actions flow from how we see the world. Our fundamental sight is what determines the paths we take, the policies we form, and the strategies we envision. One implication of this is that until we see more clearly and chart a different course of action, our policies and strategies to confront climate change are bound to fail.

The gift of sight is extraordinary. It enables us to experience the faces of those we love, beautiful sunsets, our favorite books, and work that demands concentration and skill. Our lives are enriched by this gift. But as we know, sight is not perfect. Sometimes we may misread a word, misinterpret a gaze, or see a course of action as being fruitful only to find it barren.

Our sight depends upon many factors. When we look at a sunset or read words on a page, the physical lenses of our eyes play a key role. Some people gain greater clarity with the use of glasses. Whether a person is nearsighted or farsighted, glasses can help correct that person's vision and bring clarity to replace confusion.

In other realms of sight, a mental lens associated with a worldview can influence our vision. This cultural lens is composed of our shared cultural knowledge (our shared collective values and beliefs). Just like glasses enable us to see more clearly, this lens can clarify our vision in the complex and muddled world in which we live. One critical difference, however, between glasses and this cultural lens is that while we know how glasses affect our vision, we are typically unaware of how this mental lens does so. In contrast to glasses, which enable us to more accurately see the world, a cultural lens often lets us gain clarity at the expense of accuracy. The main purpose of this book is to articulate how the cultural lens that characterizes the United States affects our vision, illustrate how it distorts our view, and demonstrate how it has led to climate change.

In other words, this book is about imperfect vision. It is about how our collective vision, or cultural lens, has led us down a path where we now face an environmental challenge that threatens our very existence. Most importantly, it is about how seeing more clearly may enable us to set a course of action that enhances the extent to which our social system is in harmony with the natural systems of the world. It is not just about seeing more clearly but also about remapping our relationship to nature, so that we can set a different course of action that will be fruitful.

Let me elaborate on how our sight, or *collective vision*, can be erroneous. At first, this may seem nonsensical. What does the concept of "collective sight" even mean? After all, isn't our sight much like a camera that takes a picture? My junior high school teacher used to say when we were having

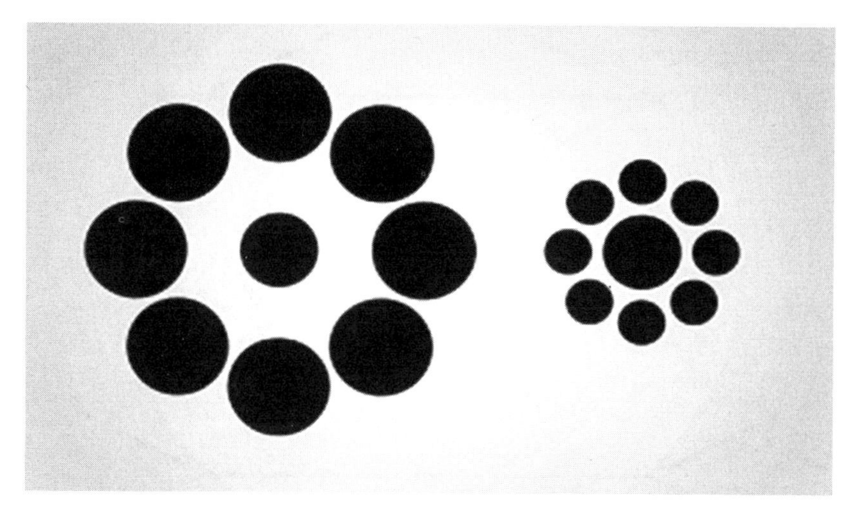

Figure 0.1. Ebbinghaus illusion, or Titchener circles. Popularized in E. B. Titchener, *Experimental Psychology* (New York: Macmillan, 1901).

class pictures taken, "The camera only shows what's in front of it." The camera doesn't distort but only records what's there.

But do our minds operate like a camera, simply recording what's there, giving rise to impressions that are accurate and true? The answer is no. When we consider ways that our sight might be inaccurate, we find there are many instances when a group of people may experience the same inaccurate vision. For instance, consider figure 0.1. Most everyone who looks at this picture sees the spots contained within the two rings as being different in size. It's obvious. If you weren't tipped off that there is probably an inaccuracy here, you might not give it a second thought. However, if you measure the size of the two spots, you'll see that your first impression was wrong. Show the picture to others and ask them what they see. As you can imagine, they'll see the same thing you did, and in that shared inaccuracy you will see how "collective sight," or a cultural lens, can produce an error.

Or, consider the line drawing in figure 0.2. Most, if not all, of us see the bottom line as being longer than the top line. This seems like a simple judgment task with little, if any, ambiguity: the difference is clearly presented to us. However, if you physically measure the lines, you'll see that they are the same length.

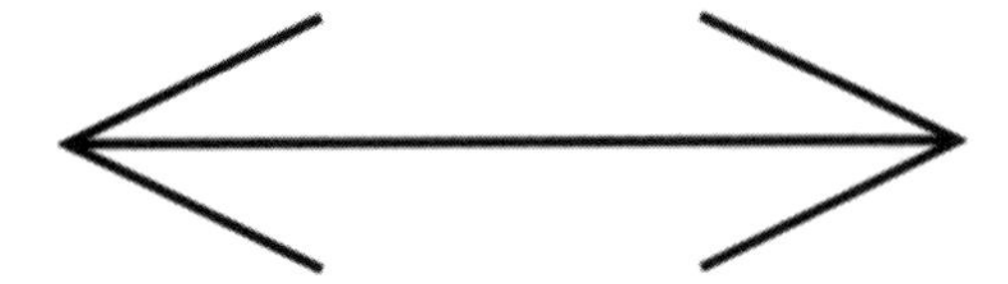

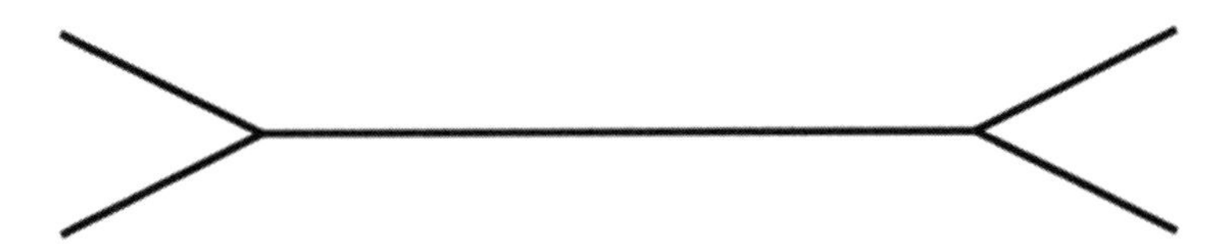

Figure 0.2. The Muller-Lyer illusion. From F. C. Muller-Lyer, "Optische Urteilstauschungen," *Archiv fur Physiologie Suppl.* (1889): 263–270.

This example is particularly interesting because psychological research points out that not every cultural group experiences this illusion (Segall, Campbell, and Herskovits 1966). In particular, people growing up in a world with rooms, where walls meet and right angles are formed, are far more likely to experience this illusion than are people not raised in a "carpentered world." So a typical resident of the United States would be far more susceptible to this illusion than a member of a hunter-gatherer group, like the San tribe of southern Africa. This example nicely illustrates that our sight is influenced by more than simply the physical lenses in our eyes, and that visual distortions are brought about by more than just shortcoming of our biologically based visual system. On the contrary, our cultural lens can cause us to perceive visual distortions. The nature of the distortion is interesting, however. Similar to the lenses of eyeglasses, this cultural lens clarifies our vision; but in contrast to corrective glasses, the greater clarity that appears to result from this lens may not always be accurate. When different cultural groups with different cultural lenses come to see the world in fundamentally different ways, this can also lead to disagreements, anger, war, and people left simply scratching their heads. An answer to a shared problem may seem obvious, and the disagreements can lead one group to consider the other group infuriatingly obstinate or even stupid.

We should stop for a moment and consider this last point. Visual errors are not necessarily associated with difficult tasks. Visual errors are also not necessarily associated with complex tasks where people feel uncertain and readily see how an error could have been made. Given figures 0.1 and 0.2, exactly the opposite point can be made. These examples illustrate that when we make an apparently simple assumption, we can be completely certain, never doubting what we see, but nevertheless be dead wrong. Because the issue may seem so obvious, however, we never question the judgment or a course of action that may flow from this judgment. We feel certain. As this book illustrates, this can get us into a heap of trouble.

But what does all of this have to do with the environmental challenges associated with climate change and our ability to face them? Read on and you'll find out. At this point, however, simply consider that our collective cultural lens can lead to misjudgments that we are likely unaware of, or that we may think are impossible. How this lens has led to misjudgments about our relationship to the natural world, and the problems this lens has created, are elaborated on later in the book.

To start, chapter 1 discusses how misperception can lead to tragedy, moves to a brief overview of the threat of climate change, and then begins to introduce how psychology can help us understand the origins of and response to this threat. Chapter 2 begins by underlining how collective vision can be erroneous, then presents the individualistic worldview, especially as manifest in the United States, and lastly, critiques it, highlighting how it has distorted our vision relative to our relationship to the natural world and can be viewed as fundamentally causing climate change. While chapter 2 focuses on how the worldview of the United States can be viewed as causing climate change, chapter 3 considers psychological obstacles that prevent people from engaging in action to remedy this problem. Shifting from the psychology behind climate change and the obstacles that impede action, chapter 4 discusses a different worldview, or cultural lens, that would lead to greater harmony between human and environmental systems, along with factors that might encourage people to transition to this different worldview. Lastly, chapter 5 illustrates that this transition is actually taking place.

In writing the book, I have tried to be neither optimistic nor pessimistic. There is a task before us, and we need to get on with it. Much like going to the doctor and receiving a diagnosis, this book is not about "happy" stories or "sad" stories. Rather, it is very much about appraising our present situation, considering the causes and obstacles to change, recommending a new perspective, and then illustrating that this change is both possible and under way. As for the basis upon which these critiques and recommendations are made, the book offers a science-based approach, arguing that testing ideas and views is the best way to correct vision as it relates to worldviews. Much like you can measure the size of the two spots in the first figure, or the length of the lines in the second, science enables us to measure aspects of our world and critique our worldviews. In this sense, science will serve as a corrective lens to help us see more clearly, chart a different course of action, and remap our relationship to the natural world.

CHAPTER ONE

See Better

Psychology as a Foundational Science to Confront Climate Change

William Shakespeare's *King Lear* is our starting point, presenting a major theme that runs throughout this book. The theme concerns sight, the accuracy of our views, and the importance of seeing clearly, for a basic premise of this book is that our distorted vision has given rise to climate change. One implication of this is that if we are to successfully navigate the environmental threat of climate change that lies ahead of us, we need to correct this distorted sight. Only in this way can we set a new course of action that will effectively confront this threat.

In Shakespeare's play, King Lear did not see clearly, and his misrepresentation of reality cost him dearly. A pivotal point in the play occurs when Lear asks his three daughters to profess their love for him. Two of his daughters, Goneril and Regan, have no particular love for their father, but simply say the words he wishes to hear. The third daughter, Cordelia, genuinely loves him in her own way, but finds herself unable to engage in this public stunt. Lear, an old man who wishes to have his ego fed by the pronouncements of his daughters, misperceives the situation. He praises Goneril and Regan for their pronouncements and angrily denounces Cordelia, taking away her inheritance and banishing her from his kingdom.

From this misperception one subplot in the play unfolds. The king, once the highest of the high, becomes a lowly wanderer in the wilderness, perplexed, lost both physically and emotionally, and heartbroken. The tragedy is that all this could have been avoided had Lear been able, at the urging of his loyal attendant Kent, to see better. But Lear, owing to his own vanity and the veiled and forthright actions of those around him, could not accurately see what was before him. Consequently, in response to a misperceived reality, he acted in a manner that led to tragedy.

As in the case of Lear, tragedies can often be avoided. For example, during the maiden voyage of the *Titanic*, the ship's captain, despite warnings of icebergs in the vicinity, continued to proceed at nearly full speed. Whether this lack of caution was due to a sense of invulnerability, a need to keep on schedule, or failed communications, is unclear. What is clear, however, is that the captain, Edward Smith, in an interview before the ship left port, reflected that he could not "imagine any condition which would cause a ship to founder. Modern shipbuilding has gone beyond that" (Wight 2012). A clear misperception, and the tragedy of the sinking is that the deaths of the more than 1,500 passengers who perished, out of 2,224 passengers on board, might have been avoided had the threat been acknowledged.

In general, as you'll see in subsequent chapters, we often misperceive situations, and at times these misperceptions have dire consequences. Vanity and a misjudged sense of invulnerability are just two factors that have been shown to lead to misperceptions. Hopefully, by articulating these biases, acknowledging them, and taking them into account as we confront the challenge of climate change, we'll be better able to face this challenge, see it for what it is, and effectively take action.

AVERTING A PRESENT-DAY TRAGEDY

Climate change and the environmental challenges that stem from it are the potential tragedies we need to clearly see and confront today. The point of this book is not to debate whether climate change has occurred. Within the scientific community, there is a resounding consensus that this threat is present, affects our lives, and must be addressed before it worsens. Many articles and books are devoted to the science behind it (e.g., IPCC 2014; Hansen 2010; McKibben 2010). Given this, there is no need for me to reiterate this information. Rather, I'll simply introduce you to several challenges related to this threat and provide a few implications associated with it.

Climate change refers to what many people call global warming, climate destabilization, or climate chaos. The terms are meant to characterize the general warming that has been recorded worldwide and the increasingly erratic and unpredictable nature of the climate. Given that people generally think of this issue in terms of climate change or global warming, let's consider *climate destabilization* and *climate chaos* for a moment. These terms highlight the idea that weather events are more commonly occurring in places where they have infrequently occurred before, such as tornadoes in North Carolina and hurricanes increasingly affecting the northeast. Additionally, the term *climate destabilization* or *chaos* is preferred to the term *global warming*, because not every region experiences an increase in temperature at each season of the year. For instance, on the one hand, the general global-warming pattern results in hot days being hotter, more intense heat waves, more intense droughts with resultant wildfires, and milder winters for many. On the other hand, eventually, if the warmer waters from the gulf stream no longer flow north toward Great Britain owing to the snowmelt from Greenland and the Artic, Europe may experience colder winters (IPCC 2014). Overall, then, the terms *climate destabilization* and *climate chaos* capture the changing climatic conditions that are leading to more erratic and intense weather events. Increasingly, this is the nature of the climate and weather that we presently face and will be dealing with in the future. When hearing the term *climate change*, then, keep in mind its varied impacts.

Another critical point to reflect upon is that *climate change has already occurred*. It is present, affecting our lives today. We cannot somehow avoid it. We cannot make it go away. Our actions have changed the atmosphere of the earth we live on, and this change—even if we stopped using all fossil fuels today, no longer emitting even a single iota of CO_2 into the atmosphere—will remain for generations to come.

Bill McKibben (2010) states that we now live on a changed world, a world he calls "Eaarth." Human emissions of greenhouse gases have already raised Earth's temperature by .8 degrees Celsius. If we were to completely eliminate greenhouse gas emissions today, Earth's temperature will still rise an estimated additional .8 degrees Celsius because of the lag time between emissions and temperature rise (McKibben 2012). In other words, we have already raised Earth's temperature by approximately 1.6 degrees Celsius, and these changes will last for centuries. Our lives and the lives of future generations will be spent on this new planet, Eaarth.

Other scientists (Crutzen and Stoermer 2000), using different terminology, argue that we have transitioned from the Holocene period to the Anthropocene period, a name reflecting the impact that humans have had on the climate (*anthropo-* meaning "human," *-cene* meaning "new"). This transition is important to acknowledge because civilization emerged during the Holocene period, a time of a relatively benevolent climate that enabled humankind to flourish. A major question is whether human civilization can still flourish during this new period.

Generally, scientists agree that, if civilization is going to continue to prosper, it is imperative that Earth warms by no more than 2 degrees Celsius (McKibben 2012b). After that point, it will be increasingly likely that our ability to control the climate will become negligible. At some point after a 2-degree-Celsius increase, climate change may spiral out of control. All of our wants, wishes, and actions will be unable to rein in a climate that will have become uncontrollable. This "tipping point" concept is analogous to a person pushing a precariously perched boulder down a slope. Once it starts rolling, there's nothing that anyone can do to stop it. We have to do everything possible to avoid pushing our climate into a runaway scenario, for at that point civilization will be threatened. So, when thinking of this issue, *the best we can hope for is to avoid the worst that this challenge has to offer, minimizing the difference between our old Earth and Eaarth, which we now live on.*

This issue in no way means that we need not continue to focus on environmental sustainability. Sustainability is still the number one goal. Equally important, however, is the need for us to figure out how we are going to live on this new Eaarth. This issue is reflected in the increased discussion of resiliency in the environmental literature (Dodman, Ayers, and Hug 2009; Doherty and Clayton 2011; Reser and Swim 2011). Increasing the resiliency of crops to drought, and the resiliency of individuals and communities to the impact of the intensification of weather, has become the topic of that discussion. In this book, I examine the factors that contribute to personal and community resiliency.

The heat waves we experience today are like none that humans have experienced for generations. This is an example of how Eaarth is different from our old Earth. In fact, the intensity of the heat wave that gripped Europe in the summer of 2003 had not been experienced in over five hundred years, and the consequences were devastating. More than fifty-two thousand people, many of them elderly, died that summer from heat-related causes. At the peak of this heat wave, in August, over two hundred people died each day in France. Moreover, the parched landscape fueled brush and forest fires, streams and rivers ran dry, and food crops withered (Larsen 2006).

Similarly, Russia had not experienced a heat wave like it did in the summer of 2010 in over 130 years. More than 550 wildfires burned forests, grasslands, and wheat fields over some 430,000 acres. Army units were called into action to assist local fire departments in quelling these fires. Crops were damaged and destroyed by the extreme heat and by fire. A nation that had been a grain exporter placed a ban on exports (Brown 2010). This is one snapshot of Eaarth, on which we now live. The possible future in which these extreme heat waves become the norm and not the exception would be a tragedy. Indeed, these by-products of climate change could cost lives, damage the ecosystems in which we live, and jeopardize our food security.

On Eaarth today, not only are episodic heat waves more extreme, but also droughts last longer. The American plains states and the Southwest have been especially hard hit by drought. For instance, in 2011, Texas had the driest year in over 100 years (the driest since 1895, to be exact), and in 2013 California had the driest year on record. In fact, the western United States has been in a drought since the year 2000 (Rice 2014).

With increased droughts, farmers and ranchers in the West have few options but to become increasingly reliant on reservoirs of underground water. These underground reservoirs are called aquifers. The major aquifer in the Great Plains is called the Ogallala Aquifer. One of the world's largest aquifers, the Ogallala Aquifer is located under portions of eight states (Texas, Oklahoma, Kansas, Nebraska, New Mexico, Colorado, Wyoming, and South Dakota). Many farmers and ranchers in these states rely on this water for irrigation of their crops and for cattle and other animals. Eighty-two percent of the people living over this aquifer rely on it for drinking water. This reservoir is being used up faster than it is being replenished. Drought only accelerates this problem. In fact, there are estimates that this aquifer will be dry within 25 years (*BBC News* 2003). This constitutes a major challenge not only to the livelihoods of farmers in this region but also to the general habitability of the region. This is a food security issue, too. We need to avoid the tragedy of the United States' breadbasket turning into a dust bowl, of megadroughts, lasting from 30 to 100 years, turning agricultural lands and ranchlands into parched deserts. In fact, megadroughts are projected for the Southwest if we do not change our present course of action (McIntee 2015).

On a related note, warming trends also lead to less snowfall in mountain regions of the world, affecting what some refer to as "reservoirs in the sky." During summer, many farmers rely on snowmelt for irrigation. Ranchers rely on this for their animals. Others rely on it for drinking water. For instance, on April 1, 2015, the snowpack reading in the Sierra Nevada was the lowest it had been in sixty years. This led the governor of California, Jerry Brown, to announce an executive order. To highlight the ongoing drought that California is facing, Brown made his announcement from a high meadow in the Sierra Nevada, beginning his speech by saying, "We're standing on dry grass. We should be standing on five feet of snow." He went on to order towns and cities across California to cut water use by 25 percent. Emphasizing that the drought might very well persist, he acknowledged that "it's a different world" we live in today, and that, accordingly, "we have to act differently" (Megerian, Stevens, and Boxall 2015).

This issue of reductions in the snowpack is far reaching. It certainly relates to the snowmelt from the Sierra Nevada that nourishes the farmlands in the Central Valley of California and towns and cities throughout the state. Similarly, the snowmelt in the Rockies feeds the Colorado River. Snowmelt from the Hindu Kush, Pamir, and Tien Shan mountains provides water to many countries in Central Asia (e.g., Uzbekistan, Turkmenistan, Kyrgyzstan, Tajikistan, and Afghanistan). And snowmelt from the Himalayas feeds every major river in Asia, where half of the people in the world live.

Himalayan snowmelt feeds the Yellow River. Having less water in the Yellow River will directly affect China's wheat harvest. And if reduced snowmelt lowers the level of the Yangtze River, China's rice production will be directly affected. Similarly, if the water in the Ganges and the Indus Rivers is reduced, shortfalls in India's wheat harvest will occur, while if the Mekong River receives less water, Vietnam's rice harvest will suffer. In this Asian region that is so heavily populated and projected to have a dramatic rise in population in the coming years, food shortages would lead not only to human suffering but also to possible political instability and conflict. This is where we are heading as a human race if we don't curb climate change.

With the melting of the arctic ice, the Greenland glaciers, and the general warming of the oceans, sea-level rise will increasingly affect low-lying coastal areas. Already, some seaside communities on the Eastern Seaboard, ranging from north of Boston south to Cape Hatteras in North Carolina, are experiencing the initial challenge of sea-level rise (Sallenger, Doran, and Howd 2012). Many coastal cities in the United States are expected to experience ill effects from rising sea levels later in this century (*New York Times* 2016). For instance, in Boston, a five-foot sea-level rise would result in parts of Logan Airport disappearing. Boston Harbor would begin to infringe on the downtown area, and the Charles River would flood much of southern Cambridge. As for Charlestown, South Carolina, the coastline is projected to move several miles inland. A similar fate is in store for the Miami area, where the sea is expected to submerge the barrier islands, Miami Beach, and much of suburban Miami. In New York, given a five-foot sea-level rise, La Guardia Airport would be threatened by the encroachment of the East River and the port complexes would be flooded. Much of Atlantic City, too, would be flooded, as would the Meadowlands. I could go on and on, but I'm sure you get the idea. The implications of this are staggering. If we do not act to prevent such damage by limiting sea-level rise, tough decisions will have to be made. Do we spend vast sums of money to protect these areas—the facilities, the airports, and the people whose homes are located in these areas—or at some point do we simply decide that these areas are no longer habitable? And if the latter decision is made, where do these people go?

Of course, this issue affects communities and nations around the globe. Millions upon millions of individuals live in coastal areas. In southeast Asia, Bangladesh, a country of 156 million people, experiences flooding each year, which covers a quarter of the country. Climate change is making flooding worse. A number of small islands, such as the Maldives, Tuvalu, and Tegua, are being threatened by sea-level rise. Many of the residents of these islands have already packed up their belongings and relocated (*The Guardian* 2009).

As time passes in this century, climate change will increasingly force people to relocate. In 2008 alone, some 20 million people were displaced by climate-related natural disasters (*The Guardian* 2009). As President Mohamed

Nasheed of the Maldives stated in testimony to the Environmental Justice Foundation, the people in his country do not want to "trade a paradise for a climate refugee camp." Yet, if climate change goes unabated, over the next forty years an estimated 150 million climate refugees will be forced to move to other countries. This is a tragedy on a global level. We need to see that this threat never becomes a reality.

Climate change has a myriad of problems associated with it. I've touched upon only a few. Other issues include health concerns associated with tropical diseases moving farther north with the warming of the planet. For example, mosquitos that carry malaria and dengue fever are moving north as the climate warms. Warmer summers and milder winters have also led to an outbreak of mountain pine beetles, which have devastated millions of acres of ponderosa and lodgepole pine forests in the western United States and Canada. Not only does the devastation of these forests negatively affect the removal of CO_2 from the atmosphere, but also the presence of millions of acres of dead trees exacerbates the threat of forest fires. The uptake of CO_2 by the oceans has led to increased acidification of these waters and the undermining of marine ecosystems, such as the bleaching of coral reefs.

This is Eaarth, on which we now live, and these projections tell us about Eaarth of the very near future. In response, actions are being taken. Water conservation measures in California are one action. Efforts to limit the construction of new CO_2-emitting coal plants are another. Miami is taking steps to deal with flooding caused by sea-level rise. These and similar efforts are important and necessary steps to take, but they are limited. Let me explain why.

In an interesting discussion of cures versus symptom relief, Martin Seligman (2011) says that there are two kinds of medications: cosmetic drugs and curative drugs. A curative drug, such as an antibiotic, kills the bacterial invaders. With the pathogens dead, the person can stop taking the antibiotic without fear of the disease recurring. On the other hand, a cosmetic drug removes the symptoms, providing relief, but doesn't cure the disease. For instance, when treating malaria, taking quinine results in temporary relief, or suppression of the symptoms. However, if a person stops taking quinine the symptoms return. While this relief is important, it is not a cure.

When confronting the "illness" of climate change, I view many of the present interventions as more cosmetic than curative. Think, for a moment, of CO_2 emissions and overconsumption as symptoms. Interventions aimed at suppressing these symptoms are important, then, since they provide relief from these ever-increasing threats. But are they cures? Is a carbon tax really a cure, or an attempt to suppress this symptom? Is the attempt to have people turn off their televisions or not purchase anything for a day really a cure, or simply another attempt to suppress an unwanted behavior? Or, when considering the consequences of CO_2 buildup in the atmosphere, such as sea-level rise, is building a seawall to confront sea-level rise really a cure? We can build seawalls and spend millions on pumps to fight the flooding. Similarly, we can spend millions of dollars trying to suppress unwanted actions, such as overconsumption. These actions aimed at suppressing symptoms like overconsumption and CO_2 are meaningful actions that need to occur, but what about treating the underlying cause? Have many of our efforts in the past failed because they were directed at the symptoms and not at the underlying cause?

ENVISIONING PSYCHOLOGY AS A FOUNDATIONAL SCIENCE TO CONFRONT THE ENVIRONMENTAL THREAT OF CLIMATE CHANGE

Thinking about climate change and other environmental threats, Stuart Oskamp (2000) wrote an interesting article titled "Psychological Contributions to Achieving an Ecologically Sustainable Future for Humanity." In that article, he states, "It is essential for us to realize that [climate change and environmental threats] are *not* solely technical problems, requiring simply engineering, physics, and chemistry for their solution. There is a crucial role for the social sciences in these problems because *they are all caused by human behavior, and they can all be reversed by human behavior*" (375, emphasis in the original). More recently, Robert Gifford (2014) underlined the idea that since climate change is predominantly caused by greenhouse-gas-emitting human activities, it can largely be reduced by changing human behavior. The problem, however, is that "human behavior is the least understood aspect of the climate change system (Intergov. Panel Climate Change 2007). Thus, unfortunately, the main cause of the problem is the least understood element. Understanding behavior at the psychological level of analysis therefore is essential, given that the cumulative impact of individuals'

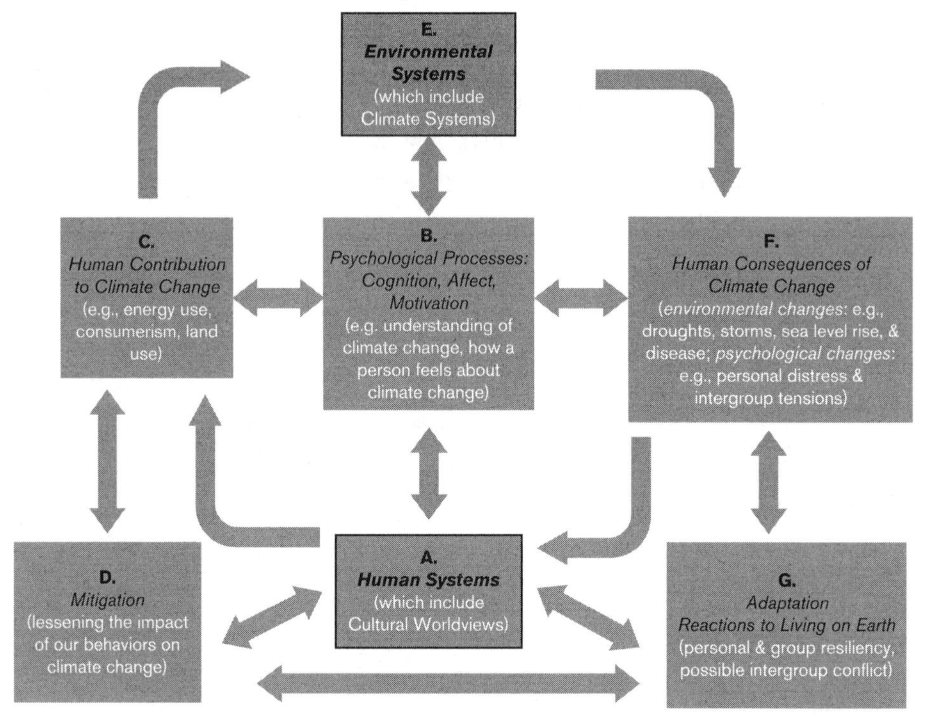

Figure 1.1. Human and psychological dimensions of climate change. Adapted from Swim et al., "Psychology's Contributions to Understanding and Addressing Global Climate Change," *American Psychologist* 4 (2011): 242, fig. 1.

decisions and behaviors is the key factor driving climate change" (Gifford 2014, p. 554). My intent in writing this book is to shed light on the psychology behind climate change.

In 2009, a summary of psychological research on climate change was presented to the American Psychological Association Task Force on the Interface Between Psychology and Global Climate Change. As a follow-up to this report, a series of articles appeared in a special issue of the *American Psychologist*. In one of these articles, titled "Psychology's Contributions to Understanding and Addressing Global Climate Change," several of the leading researchers in this area—Janet Swim, Paul Stern, Thomas Doherty, Susan Clayton, Joseph Reser, Elke Weber, Robert Gifford, and George Howard (2011)—presented an interesting model of the interplay between human and environmental systems related to climate change. I've modified this model and have used it as a framework for the contents of the present book (see figure 1.1).

Commenting on this model, I can say that the main task I see before us is to create harmony between "human systems" and "environmental systems" (the A–E relationship). In relation to climate change, this means we need to consider how, as part of the human systems, the cultural *worldview* that characterizes the United States influences "psychological processes" (i.e., thoughts, feelings, and motivations) that cause the "human contribution to climate change" that negatively impacts "environmental systems." In the model, this is the flow from A to B to C to E. I discuss this part of the model in chapter 2.

Mitigation (D in the model) refers to actions people can take to lessen their impact on the climate. There are various barriers, however, that prevent people from engaging in mitigating actions. Chapter 3 examines these barriers. One barrier concerns the direct impact that the worldview predominant in the United States has on mitigation (the A to D relation), while other obstacles relate to the impact this worldview has on other psychological processes that limit mitigation (the A to B to C to D flow). Moreover, obstacles associated with psychological processes that are independent of this worldview are presented in this chapter (the B to C to D link).

An alternative worldview that is positively associated with mitigation and reducing the human contribution to climate change is presented in chapter 4. An overview of the research illustrating the positive impact of this worldview on creating greater harmony between human systems and environmental systems is presented in this chapter. Additionally, I highlight the positive impact of this alternative worldview on other psychological processes. With reference to the model, chapter 4 explores the same links as in chapter 3, but instead of focusing on the negative impact of the U.S. worldview, my discussion in chapter 4 shifts to the positive impact of this alternative worldview. In a follow-up to this research overview, chapter 5 illustrates how communities and individuals are actually transitioning to this alternative worldview.

Lastly, chapters 4 and 5 also consider our life on this new planet, Eaarth. These chapters touch upon the A to B to F to G set of relationships. Given the "human consequences of climate change," issues concerned with how we are to adapt to living on this new planet, Eaarth, need to be addressed. Overall, this model illustrates the various issues and questions I address in this book.

The Psychology of Perception versus Reality

A basic premise of this book is that the worldview that characterizes the United States is the underlying cause of climate change. This worldview leads us to misperceive our relationship to the natural world, and the distorted view that results from this worldview has led us into an environmental trap. From this perspective, if we are to truly avoid the tragedy associated with climate change, we need to change our worldview. We need to remap or reorient ourselves, so to speak, to the way we see ourselves relative to nature and, in this remapping, to discover a new collective path to take. Before venturing further into this argument, however, let me lay a bit of groundwork.

First, I need to elaborate on the distinction between perception and reality: "see better" is not just a statement that Kent makes to Lear. From psychology's very beginning, psychologists have been interested in helping people to more clearly see themselves, their motivations, the impact that they have on others and that others have on them, and why they act the way they do. In doing so, psychologists have identified a number of factors that lead people to misperceive the world in which they live.

For instance, since early in psychology's history the distinction between perception and reality has been of paramount importance. As depicted in my adaptation of Brunswik's (1939, 1952) lens model (figure 1.2), the objective image of reality is transformed through specific constructive processes of the mind. These constructive processes include knowledge structures, principles of unit formation, and the desires of the self, or self-motives. As you can see in this figure, the unit formation principle of good continuation might lead a person to perceive the snake as being a "better form" (i.e., perceive it more clearly) than the objective reality may warrant.

Knowledge structures include our shared cultural worldview and other idiosyncratic knowledge structures. For instance, the constructive processes of the mind, as related to the shared cultural worldview regarding snakes, may lead a person to fear snakes. Generally speaking, our shared knowledge structure of snakes is one of concern, wariness, and fright. However, our idiosyncratic knowledge of snakes may also come into play. For example, I grew up around snakes. I know that certain snake species, such as the rattlesnake, are to be feared, while others, like the king snake, are

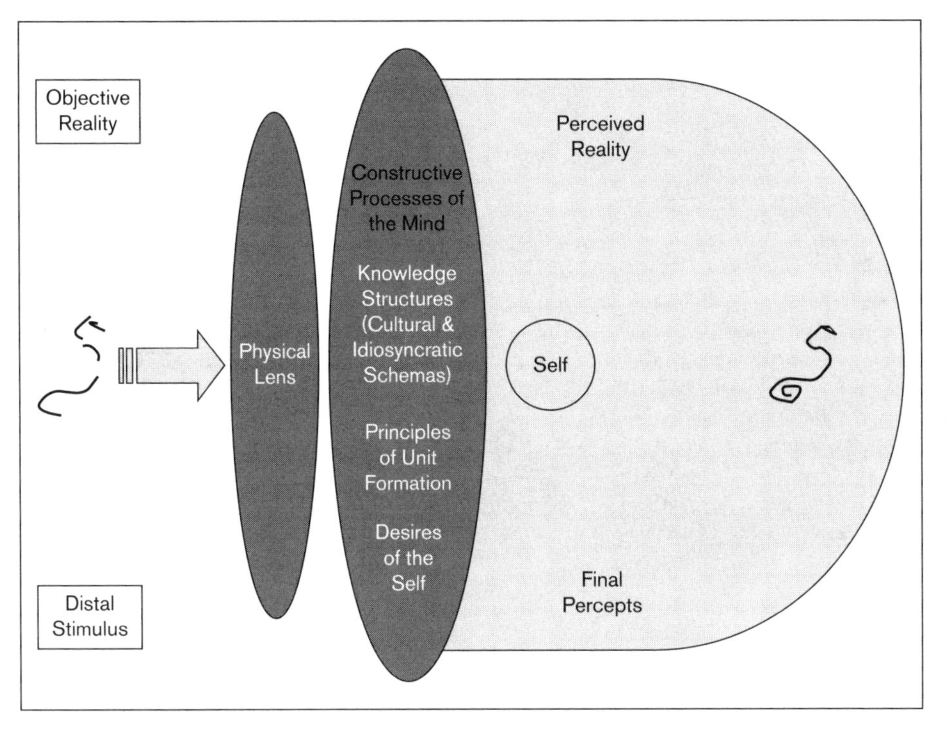

Fig 1.2. Adaptation of Brunswik's lens model. E. Brunswik, *The Conceptual Framework of Psychology*, vol. 1 of *International Encyclopedia of Unified Science*. Chicago: University of Chicago Press, 1952).

harmless. In other words, my idiosyncratic experience (i.e., my unique experience, which is generally not shared by others) may lead me to operate in a perceived reality very different from that of someone who relies only on the cultural worldview. I would be calm in the presence of a king snake. I might even approach it and pick it up. On the other hand, my wife (who did not grow up around king snakes) would not act nearly so calmly. Thus, even under identical objective realities, the perceived realities we operate within can dramatically differ from those of another person, owing to how our minds transform this objective information.

Early work on the constructive processes of the mind emphasized how experiential factors can serve as part of the constructive process of the lens. The constructive process transforms the objective reality into a perceived reality that is often very different. The distinction, then, between the distal

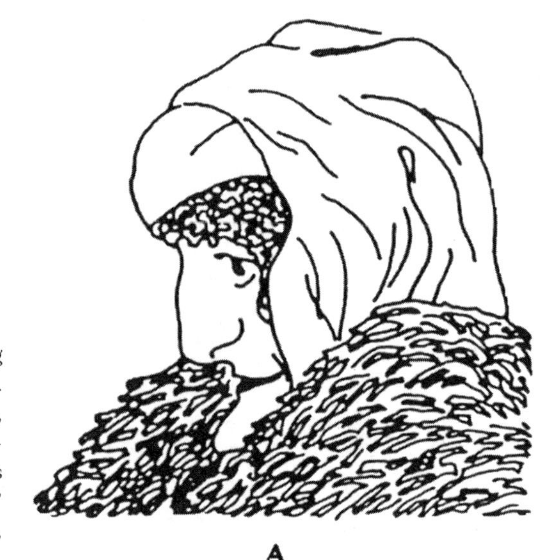

Fig 1.3. Old woman / young woman illusion, by W. E. Hill. The original caption read, "My wife and my mother-in-law. They are both in this picture—Find them." *Puck* magazine, November 6, 1915, p. 11.

Fig 1.4. Old woman / young woman illusion, by W. E. Hill. The original caption read, "My wife and my mother-in-law. They are both in this picture—Find them." *Puck* magazine, November 6, 1915, p. 11.

stimulus and the final percepts is critical, for what we "see" may greatly differ from what is objectively "out there."

For an illustration of this point, look at figure 1.3 and then figure 1.4. What do you see? Most people see the second image as consistent with the first image (i.e., as an old woman). Now show figure 1.5 to someone and then figure 1.4. What do they report seeing? Most people who view figure 1.5 first

Fig 1.5. Old woman / young woman illusion, by W. E. Hill. The original caption read, "My wife and my mother-in-law. They are both in this picture—Find them." *Puck* magazine, November 6, 1915, p. 11.

will see figure 1.4 as the image of a young woman. Think about this for a moment. When presented with identical images (i.e., figure 1.4), people more often than not see them differently because of their prior experiences (i.e., depending upon whether they first saw 1.3. or 1.5). Their prior experiences lead to expectations, and these expectations lead them to "see," not the reality of an ambiguous figure, but an articulated image consistent with their expectations.

Schemas and Misperceptions

In Brunswik's day, one's initial exposures to an image were thought to create a mental set, a cognitive framework that became part of the constructive process of the mind, which predisposed people to view an ambiguous image in a manner consistent with the mental set. Today, mental sets are called schemas. Schemas are defined as organized bodies of knowledge. In the young woman / old woman example, the organized body of knowledge was the unambiguous old or young figure to which a person was initially exposed.

For a real-life example, think of the city of Los Angeles for a moment. We all have an organized set of beliefs and attitudes about L.A. For myself, congested freeways, urban sprawl, sunny beaches, the Lakers, and Hollywood are ideas that easily come to mind. For others, a different cluster of ideas may be present. The main point is that an experience is organized. Whether the experience is about L.A., the high school we went to, our mother or father, vegetables, birds, or a romantic partner, this organized body of knowledge, or schematic representation, is part of the constructive process of the mind that influences our perceptions of what we see.

More specifically, schemas influence the way we see the world by making *information consistent with the schema easier to remember and retrieve* from memory. For instance, my schema of L.A. makes it very easy for me to retrieve memories of congested freeways, beautiful beaches, and so on. Additionally, however, schemas *lead us to anticipate events and act on the basis of what we anticipate*. For example, given that one aspect of my schematic representation of Los Angeles is that I anticipate freeway traffic, when I fly into L.A. I try to arrive at a time when I expect less traffic to be present, such as midmorning on a Sunday. Moreover, schemas lead us to *selectively expose* ourselves to only certain information. For instance, given that I have an overall favorable impression of L.A., I am more likely to seek out information that confirms this view rather than information that disconfirms it.

However accurate or inaccurate individuals' perceptions of the world may be, owing to their schematic representations, another main function of schemas is that this lens aids people in fulfilling their need for understanding. Schemas add coherence and continuity to individuals' and groups' everyday experiences. In a world of complexity and inherent ambiguity, it is impossible to pay attention to everything. Schemas help us anticipate and simplify this hum and buzz of reality, making it more manageable. Of course, the price we pay for this psychological lens that adds clarity and continuity to our experience is that these clearer perceptions may be clearly inaccurate.

We might think of this disjunction between perception and reality, in its most extreme form, in terms of the person who has lost touch with reality. In some instances, individuals may erroneously see others stalking them and may be afraid to leave their residences, or they may perceive themselves to be so special that the rules that apply to others have no bearing on them. In these extreme forms, we talk of psychopathology and narcissism, respectively. However, what psychological research points out is that to a lesser extent, we all are prone to distort our view of the world in one way or another.

For instance, in order to protect our self-esteem, we might not take responsibility for some negative event for which we were largely responsible. Instead, we might provide an excuse, blaming circumstances or someone else for what happened. Or, in order to enhance our sense of self, we might take more credit for some positive event than we actually deserve, basking in a warm glow of positive feelings that accompanies the success. As you progress through this book, you'll encounter a myriad of factors that lead individuals to misperceive the world around them.

Cultural Worldviews, Cultural Schemas, and Unjustified Clarity Distortions

Based on their unique personal experiences, individuals differ from one another in terms of the schemas they possess. We each have a personal history with a set of experiences that never completely overlaps with the set of experiences of another person. On the other hand, we each also grow up in a specific cultural context, and the shared experiences that we have in this context can lead to a shared cultural schema. Before I develop this idea in more detail, however, let's examine how psychologists define culture and what elements tend to comprise cultural schemas.

Heine (2015) defines a culture as a particular group that shares a particular kind of information that distinguishes it from other groups. This information (ideas, beliefs, habits, or practices) is acquired through social learning. A similar definition is presented by Matsumoto and Juang (2012, p. 15), but they also emphasize that this information is "transmitted across generations" and "allows the group to meet basic needs of survival, pursue happiness and well-being, and derive meaning from life." This definition highlights the idea that this shared information is lasting and helps fulfill needs of the group.

Triandis's (1996) definition underscores the idea that this shared information includes "shared attitudes, beliefs, categorizations, self-definitions, norms, role definitions and values," meaning, among other things, that cultures lead us to think of ourselves in certain ways, lead us to have shared expectations about how people should or ought to act and to consider the appropriate roles that women and men can perform in their social lives. The value concept underlines these "should" and "ought" statements. Values are concepts, or beliefs, that pertain to desirable behavior directed at particular goals. They transcend specific situations, guide selection and evaluation of behavior or events, and are ordered by relative importance. In his view, then, cultures are far more than simply a particular group sharing particular information. Cultures influence what we strive for in life and how we evaluate our own and other's behaviors. Moreover, his definition ends by stressing that this information is all "organized around a theme," which emphasizes that a central idea unites this information and gives coherence to it.

In his reflections on the meaning of culture, Jerome Bruner ties these ideas to the notion of a folk psychology, or commonsense psychology, which reflects how people in a particular culture typically think. Elaborating on this idea, Bruner (1990) states,

> All cultures have as one of their most powerful constitutive instruments a folk psychology, a set of more or less connected, more or less normative descriptions about how human beings "tick," what our own and other minds are like, what one can expect situated action to be like, what are possible modes of life, how one commits oneself to them, and so on. We learn our culture's folk psychology early, learn it as we learn to use the very language we acquire and to conduct the interpersonal transactions required incommunal life. . . . Folk psychology is about human agents doing things on the basis of their beliefs and desires, striving for goals, meeting obstacles which they best or which best them, all of this extended over time. (pp. 35, 42–43)

As we can see, then, culture not only is external to a person and exerts pressure on a person to believe and act in certain ways but also becomes internalized by members of a culture. The internalization of a common set of values and beliefs, organized in the minds of the members of a particular culture, leads them to see things and respond in ways that are characteristic of that culture. That this information is organized in the minds of members of a given culture has led Markus, Kitayama, and Heiman (1996) to relate this concept to people having shared cultural schemas. As they state, "Cultures provide the interpretative frameworks—including the images, concepts, and narrative, as well as the means, practices, and patterns of behavior—by which people make sense (i.e., lend meaning, coherence, and structure to their ongoing experience) and organize actions. Although experienced as such, these organizing frameworks (*also called cultural schemas*, models, designs for living, lifeways, modes of being) are not fully private and personal; they are shared" (p. 858, emphasis added).

When we relate the concept of cultural schemas back to Brunswik's lens model, we see that part of the constructive aspect of the mind that comes between objective reality and perceived reality, and which can lead the perceived reality to differ from the objective reality, involves a person's cultural schema, or cultural lens. Like a physical lens associated with a pair of glasses that helps us see better, this cultural schema, or lens, can also lend coherence and help articulate our everyday experiences. Our day-to-day life seems clearer. Clearer vision, though—like the clearer vision illustrated in the old woman / young woman example, where the objective reality of an ambiguous figure is transformed into either an unambiguous old or an unambiguous young figure—doesn't necessarily mean more accurate vision.

This is an interesting meaning of *distortion*. Typically, when I think of distortion, such as when listening to music, I think of a clearer sound becoming less clear. Or, when considering visual distortion, I think of how a painting may seem blurry and how my glasses help correct this. In contrast, in order to understand my usage of the word *distortion*, think of how the mind might transform "noisy" music so that one can hear a clear melody. Or, when looking at a painting with multiple potential meanings, think of how the mind might lead a person to see only one of them. In other words, clarity where clarity is not present is also a distortion.

So, instead of thinking of how a pair of glasses may lead to a more accurate representation of reality, consider the opposite possibility. Think of the implications of what I call an *unjustified clarity distortion*. One major implication is that by *not* recognizing the ambiguity, a person may have confidence in a perspective that is unwarranted. Given the unjustified clarity of what a person may see, it may be easy for the person to judge an alternative perspective as being wrong.

Do Cultural Schemas Help or Hinder Us?

When we consider how cultural schemas or a cultural worldview can distort reality, the next question becomes: In what ways and to what extent does this occur? Additionally, when thinking of these distortions, the question of whether these distortions help or hinder us comes into play. Recall Matsumoto and Juang's definition of culture: a worldview is supposed to help "the group to meet basic needs of survival, pursue happiness and well-being, and derive meaning from life." Does it necessarily do so, though?

Jared Diamond (2005), in his book *Collapse*, illustrates that a cultural worldview can be inaccurate and lead to the downfall of a group of people. Providing one example from his book, he recounts the fall of the culture that existed on Easter Island. Population growth combined with the elimination of trees on the island proved deadly for them. The deforestation, however, made sense within their worldview. They believed that their gods protected them, and this led them to chisel huge stone statues of their gods and to transport these statues to different points on their island. Trees were used for this transport. Deforestation, however, undermined their food source, for the trees had provided food (palm hearts), habitat for wild animals, fuel for heat and cooking, wood for fishing canoes, and roots to prevent erosion of the soil. Thus, while their worldview provided them with clarity and meaning, and they anticipated from their set of beliefs that their actions would lead to their well-being and safety, the inaccuracy of this shared cultural representation undermined their existence. Their unjustified clarity distortion, which in all likelihood led them to feel very certain and justified in their actions, led to the collapse of their culture.

Diamond presents eight processes through which past societies have damaged their environments and undermined their existence (deforestation and habitat destruction; soil problems; water management problems; overhunting; overfishing; effects of introduced species on native species; human population growth; increased per-capita impact of people). Typically, these are the usual suspects that environmentalists think about when considering whether we are undermining our existence. These discussions often involve whether modern farming practices, with their heavy use of chemical fertilizers, are actually harming the soil. How our encroachment on the natural world through industrial/business practices and housing development projects may lead to a loss of habitat that negatively affects our food chain. Whether our overuse of water from aquifers is sustainable, and whether our carbon footprint, which relates to population size and consumption practices, is sustainable. These are all critical questions, and we touched upon several of them earlier in this chapter.

Interestingly enough, however, Diamond also proposes psychological factors related to faulty group-decision-making that contribute to societal collapse. Among them, he considers how groups may fail to "anticipate a problem before the problem actually arrives" (2005, p. 421). He also considers how a society may collapse because the problem arrives and people fail to see it. One other factor especially important for our discussion involves failure that occurs when a group sees the problem, tries to solve it, but can't. Additionally, he emphasizes the importance of intergroup relations in the form of hostile neighbors or the lack of friendly trade partners.

These factors framed within our discussion of cultural groups and shared cultural schemas are highly related to the impact that schemas have on our information processing. Recall, from our earlier discussion of schemas, that the latter involve the organization of information and *anticipation* of events. That is, schemas lead us to anticipate the occurrence of certain things and not others. Consequently, we are more likely to look for and actually see the anticipated event, often ignoring other events. Diamond's proposed factors need to be examined from a cultural worldview or schema perspective. Specifically, in subsequent chapters I address how the worldview that characterizes the United States has hindered our anticipation of the climate change problem. Additionally, I address how this deters people from seeing the effects of climate change that have already occurred around them. Lastly, I tackle the question of whether a worldview that caused the climate change threat can potentially solve it.

Generally speaking, we need to consider whether the cultural worldview that characterizes the United States distorts reality. Does it help us survive? Might it lead to societal collapse? Does it promote our well-being and happiness? Is there a better, alternative worldview? These are some of the critical issues and big questions I address in this book.

The Issue of Critiquing Worldviews

It may seem odd or presumptuous of a person to critique a worldview. After all, who's to say what is correct? Science can help us here. We can empirically test ideas and sort out truth from fiction. Scientific observations can lead us to develop new and more accurate worldviews.

For instance, think about how scientists critiqued the worldview that the earth was flat and that the sun revolved around the earth. When people who believed in this worldview woke up in the morning and walked outside, it probably seemed obvious to them that the world was flat and the sun traversed the sky overhead. But, given how schemas can lead to misperceptions, the impact of the schematic representation of these flat-Earthers and believers in geocentricism may have led them to seek, more easily encode, and remember information consistent with their representation. Stated differently: they may not have been open to other perspectives. They may have judged other perspectives as incorrect. However, as scientific evidence mounted over time, it became clear that this worldview was wrong and that Copernicus's heliocentric view was correct.

As for psychologists critiquing cultural worldviews, I am certainly not the first. From the earliest writings of Freud ([1927] 1989), in which he considered the "future of an illusion," to Fromm's (1955) book *The Sane Society*, where he considered how societies could be more or less sane, the work in psychology has been, not just about how individuals misperceive the world around them, but about how groups do so, as well. Fromm states that members of cultures can be socialized to develop "socially patterned defects" (p. 15), and he elaborates on how these defects can affect the quality of our lives. *What is unique about the present book, in contrast to this earlier work, is that it articulates how socially patterned defects that result from incorporating the worldview associated with the United States can lead people to harm the natural world and threaten their very existence. Additionally, I review scientific research that examines whether this worldview promotes well-being and happiness. Moreover, in this book I present an alternative worldview that has been shown to enhance harmony between social and environmental systems, happiness, and well-being.*

Before continuing with this argument, it is important to step back for a moment and reflect on what I offer in this book. Generally speaking, I have brought empirical research to bear on shortcomings and costs associated with a worldview that many people in the United States hold near and dear. When Tim Kasser, Steve Cohn, Allen Kanner, and Richard Ryan (2007) critiqued American corporate capitalism (ACC), they made a statement at the beginning of their overview of the literature that bears repeating.

> Finally, we acknowledge that psychologists are often reluctant to explore emotionally and politically charged topics such as the effects of living under particular economic systems (Kasser and Kanner, 2004). This reluctance may be due to a fear among psychologists that they would be considered politically incorrect, unaware of cultural relativism, or even unscientific were they to explore such matters. Indeed, when we proposed the present target article, one of *Psychological Inquiry*'s editors cautioned us to write extra carefully so that readers did not "discount the ideas as left wing propaganda." We hope that our colleagues will not treat the mere fact that we are explicitly examining some costs of ACC as evidence of political bias; to do so would render the economic system a sacred cow excused from the intellectual and empirical scrutiny that science encourages, and requires, for any topic of investigation. (pp. 3–4)

Similarly, I hope that my critique of this worldview will not be viewed as anything other than a scientific appraisal of this lens that affects so many aspects of our lives.

As for the alternative worldview I present, it is based on the work of environmentalist Aldo Leopold and referred to as the "land ethic" worldview. Why this worldview rather than another one? Once again, research can help us here. For instance, I could present a worldview based on the New Environmental Paradigm (Dunlap and Van Liere 1978; Hawcroft and Milfont 2010). Since the late 1970s, this worldview and the corresponding scale that measures the extent to which individuals have adopted it has guided much of the research in this area of environmental psychology. In research I conducted with a colleague (Mayer and Frantz 2004; Frantz and Mayer 2013), however, we found that the land ethic worldview is a consistently better predictor of proenvironmental behavior and well-being than this other worldview. In fact, statistically controlling for participants' land ethic worldview scores, we have found that participants' New Environmental Paradigm scores are unrelated to these measures. In contrast, statistically controlling for participants' New Environmental Paradigm scores, we find that our measure of the land ethic worldview remains significantly related to both proenvironmental behavior and well-being scores. Consequently, on the basis of our empirical findings, I present the land ethic worldview as the alternative worldview to which we need to transition.

That the environmental challenges we face are largely due to our misperception of reality is nicely captured in a quote from the environmentalist David Orr (2009), who states, with respect to the challenge of climate change, that it "is not an aberration but a predictable outcome of a system haphazardly created in the dim light of a dangerously incomplete image of reality" (p. xii). At its very core, psychology is a science that concerns itself with individuals' and groups' perceptions and misperceptions of reality. As a result, psychology is a foundational science that can help us to confront and deal with this challenge.

The Emphasis within the Environmental Movement: Is it Severely Limited?

The emphasis within the environmental movement has been primarily on economic, political, and technological issues as a way of dealing with the climate change challenge. Economists, such as Paul Krugman, argue for a carbon tax. Politically, representatives from 195 countries worked on the Paris climate agreement. Technologically, cars that are more efficient and less polluting, lower-cost solar panels, better energy-efficient lighting options, and better-insulated homes are among a host of technological improvements aimed at reducing carbon emissions. While these are all meaningful approaches, they fail to address cultural worldviews that gave rise to the threat in the first place. In other words, in trying to remedy the climate-change crisis by treating the symptoms or behavioral practices that derive from a more fundamental worldview, the effectiveness of these interventions may be very limited.

A major implication of the argument I make in this book is that before you can realistically expect economic interventions to gain broad popular support, politicians to feel pressured by the electorate to work on and be committed to favorable environmental policy, and technological advancements to really take hold in individuals' lives, people need to see the importance of these changes. As Bill McKibben (2012a) states, "The politics are probably impossible—unless somehow we can build a movement that can really push. . . . That movement can only come when we feel, deep down, the impact of what's happening around us" (p. 15). From my perspective, this essential change can be brought about by encouraging people to adopt a worldview that fosters a more accurate representation of our relationship to the natural world: a worldview that helps us see and feel how our individual and collective actions affect that world.

In Summary: Five Major Issues

To reiterate, five major issues are covered in this book. The first issue considers the underlying psychological cause of climate change. Chapter 2 presents the argument that the worldview that characterizes the United States is the primary cause of climate change, and I highlight distortions that arise from this worldview. The links between these distortions and climate change are the primary focus of that chapter. Additionally, when evaluating this worldview, I consider whether it promotes happiness and well-being. Elaborating on the distortions of this worldview, linking it to climate change, and providing research evidence that it does not promote well-being and happiness set the stage for considering an alternative worldview that does promote our survival, well-being, and happiness.

Before articulating the alternative worldview, however, I consider, in chapter 3, why people often fail to take action to confront climate change. In other words, the *cause* of climate change is the first theme of the book, while the second theme is the wide variety of psychological processes that lead people to do nothing about it. The distinction between chapters 2 and 3 can be illustrated with a fire analogy: We can consider the cause of a fire (a lightning strike, matches, etc.). In contrast, we can also consider why people may fail to take action to put the fire out (fear of getting hurt, being unaware that the fire even existed, etc.). These are separate questions, and each of the two chapters addresses one of them. While the worldview distortion found in the United States plays a role in both themes, other psychological processes also need to be considered when focusing on the second theme.

When considering why people may fail to take action to confront climate change, my discussion highlights research on the psychology of helping. Social psychology has a rich history of research in this area and has uncovered many barriers to helping. This research, however, typically addresses why people *fail to help other people* (i.e., it centers on a lack of pro*social* behavior). In contrast, I've adapted one model of helping to illustrate how it also pertains to why people *fail to help alleviate the environmental crisis* we face (i.e., why people fail to engage in pro*environmental* behavior as related to climate change). Besides illustrating how the worldview of the United States negatively affects proenvironmental behavior, I present factors related to motivation, group processes, denial, and persuasion.

The third major issue covered in this book concerns the remedy, the land ethic alternative worldview, discussed in chapter 4. After presenting the characteristics that comprise this worldview, I turn to research illustrating how it promotes proenvironmental behavior as it relates to climate change. I also address research related to whether this alternative worldview promotes not only survival (i.e., by reducing the threat of climate change) but also happiness and well-being. Moreover, I extend this worldview to issues concerning peaceful relations between people and between groups. One issue here relates to whether we can effectively confront this environmental crisis if conflict between individuals and groups is still prevalent. Thus, this chapter deals with how we might promote greater harmony not only between human and environmental systems but between human systems as well.

The fourth issue, presented in chapter 5, considers both existing programs that are leading people to adopt this alternative worldview and how change to this alternative worldview might be promoted. In other words, the transition to this alternative worldview is already taking place. The alternative worldview is not just a research-based idea but a reality that is gaining traction in different parts of the world. Additionally, in this chapter I both document how this transition is taking place and seek to empower people by providing them with means by which they might make the transition themselves. Lastly, the fifth issue, touched upon in chapters 4 and 5, concerns the challenges of living on Eaarth and how this alternative worldview might help us face those challenges.

FINAL THOUGHTS

The Brundtland Report of the World Commission on Environment and Development (WCED 1987) defined sustainable development as "meeting the needs of the present without compromising the ability of future generations to meet their own needs" (p. 363). As Oskamp (2000) so aptly puts it, "Starkly stated, the issue is whether there will be a livable world for our descendants and other creatures to inhabit" (p. 373). Climate change stands as a major threat to our present needs and to the needs, dreams, and aspirations of those who will come after us. Addressing the issue of whether we can stand up to this threat entails standing up and facing our own worldview that has created the threat. Looking in the mirror, we have to realize that the "enemy" is largely ourselves and the behavior patterns that flow from our own representation of the world.

A sense of tragedy is not something that pervades the worldview of the United States as it did the worldview of the ancient Greeks. In our modern life we generally expect happy endings and progress. Each generation expects their standard of living to exceed the standard of living of the previous generation. The aim of this book is to provide answers and a path to a better life. As you read these pages, however, you might note an underlying tone indicating that this better life is not guaranteed. There is a ticking clock, and at some point a tipping point will be reached. If changing our worldview and overcoming barriers to proenvironmental action is the answer, then another question arises: can this change occur before the clock ticks past that tipping point? I don't have an answer to this question, and I don't think anyone does.

And while this book does present a path to what research suggests is a better place, you have to remember that this better place is on Eaarth. What does this mean for us? What does a path to a better place on a compromised world entail? How will we navigate this different path?

Some people see similarities between the history of the industrialized world, especially as exemplified in the United States, and the maiden voyage of the *Titanic*. For instance, as Ronald Wright in *A Short History of Progress* states, "Our civilization, which subsumes most of its predecessors, is like a great ship steaming at speed into the future. It travels faster, further, and more laden than any before. We may not be able to foresee every reef and hazard, but by reading her compass bearing and headway, by understanding her design, her safety record, and abilities of her crew, we can, I think, plot a wise course between the narrows and bergs looming ahead. And we must do this without delay, because there are too many shipwrecks behind us" (2004, p. 3). My goal in writing this book is to enable the reader to see the hazards in front of us as well as possible and to present a different course. It is about reading our cultural "compass bearing" and questioning our speed and the direction we're heading. It is about understanding the members of our culture, their wants, desires, abilities, and fears. It is about reorienting our strengths in a way to avoid the tragedy of a possible shipwreck and promoting the likelihood of a safe passage through very challenging times. Ultimately, it is about whether systems devised by humans, as represented by their worldviews and resultant lifestyles, can exist in harmony with the natural world.

We need to see that improper vision has led to many of the challenges we face today, and that improper vision leads to improper solutions. How we view ourselves, the meaning of our lives, and the possibilities available to us will play a critical role as we confront the environmental challenges before us. Psychology has much to offer as a science that studies our representations of the world in the form of our worldviews, our motives, factors that lead to conflict and conflict resolution, and healing. It can help us chart a fruitful path into the future and identify barriers and pitfalls that need to be avoided as we move forward, for fashioning as fruitful a life on Eaarth as possible is the ultimate goal.

CHAPTER TWO

The Fundamental Problem

The Psychology behind Climate Change

Chapter 1 introduced the idea that distorted vision can lead to tragedy, and that we need to do everything possible to avoid the potential tragedy associated with the threat of climate change. My aim in the present chapter is to articulate how the cultural lens of the United States distorts our vision and can be directly linked to the pressing issue of climate change. However, given that people typically never question the authenticity of their worldviews, it seems important to initially return to and extend the discussion of how a cultural lens can lead to distorted vision. Then we'll be able to consider how the distorted vision associated with the cultural lens of the United States can be linked with climate change.

UNJUSTIFIED CLARITY DISTORTIONS: A FURTHER ELABORATION

An ironic aspect of worldviews is that they can both clarify and distort how we see the world. In other words, a cultural group can develop a shared understanding of reality, feel certain that their views are correct, and be dead wrong. How can this happen? As demonstrated by the old woman / young woman image in chapter 1, our seeing is influenced by our prior experiences. The internal lens, or schemas, we develop predispose us to see certain things to the exclusion of other things, providing apparent clarity where clarity is not actually present, where another view might be correct. We live in a world of unjustified clarity distortions.

Our attention is selective. We simply are not capable of paying attention to every single thing. So, given this limitation, we select certain information to focus on at the expense of other information. Is this a conscious process? Sometimes it is. However, as illustrated in the old woman / young woman example, at other times our internal lens subconsciously highlights certain anticipated information to the exclusion of other information, and we simply see what we expect to see.

When a group of people located in a particular place share a common schema, it is referred to as a cultural schema, a cultural lens, or a cultural worldview. This shared cultural lens leads a group to see their world more clearly within the framework of their worldview. This mind's eye clarifies ambiguities people experience. It fulfills our basic need for understanding. The world seems less chaotic and more comforting as a consequence. However, clarity may be gained at the expense of accuracy. People may not acknowledge ambiguity where ambiguity exists. People may not be able to see alternative views. As in the old woman / young woman example, people who are predisposed to see the young woman often have difficulty seeing the old woman.

The experience of certainty in the face of ambiguity can be problematical. After all, there is no guarantee that the view is correct. In the old woman / young woman example, the reality of the situation is that the figure in the ambiguous image is neither definitively young or definitively old. Given that this image can be viewed in multiple ways, to clearly see it as only one image is a distortion. And given that many of our everyday life experiences are ambiguous and can be viewed in multiple ways, the fact that our cultural lens leads us to see each of these experiences in only one way can similarly be considered a distortion of this reality.

As an example, consider the poem *The Right Kind of People* by Edwin Markham, which like the old woman / young woman example highlights how past experiences lead us to interpret our everyday experiences in a manner consistent with our expectations.

Gone is the city, gone the day,
Yet still the story and the meaning stay:

Once where a prophet in the palm shade basked
A traveler chanced at noon to rest his miles.

"What sort of people may they be," he asked,
"In this proud city on the plains o'erspread?"
"Well, friend, what sort of people whence you came?"
"What sort?" the packman scowled; "why, knaves and fools."
"You'll find the people here the same," the wise man said.

Another stranger in the dusk drew near,
And pausing, cried "What sort of people here
In your bright city where the towers arise?"
"Well, friend, what sort of people whence you came?"
"What sort?" the pilgrim smiled,
"Good, true, and wise."
"You'll find the people here the same,"
The wise man said.

As this poem points out, we can, like the prophet in the poem, anticipate that individuals' new experiences—which are full of ambiguities and include people's new experiences of others, who can be viewed either as "knaves and fools" or as individuals who are "good, true, and wise"—will be unjustifiably clarified, "seen" in a manner consistent with people's prior experiences. In other words, what is "seen" is often more a reflection of the seer than it is of what is being seen. Or, rephrasing a common saying, instead of thinking that "seeing is believing," it may be more accurate to state that "believing leads to seeing."

Or, when thinking about how groups can engage in distorted vision, consider the cult group the Seekers, whose leader, Marian Keech, claimed to be in contact with the Omnipotent Guardians from the planet Clarion (Festinger, Riecken, and Schachter 1956). She reported obtaining messages from the Guardians via telepathy, automatic writing, and gazing into a crystal ball, and these messages informed her that humanity would be destroyed as a result of a flood of biblical proportions before dawn on December 21, 1954. She attracted a group of followers, and they developed a firm set of shared beliefs, a worldview, concerning the planet Clarion, the Omnipotent Guardians, why humanity would be destroyed for its sins, and why they alone would be spared.

This worldview clarified and sharpened their vision of their world. They knew what was right, and they knew what they had to do. Given their clear vision, several members quit their jobs and spent all of their savings before the fateful dawn of December 21. Of course, their worldview was incorrect. The flying saucer never arrived, the flood never occurred. Their clear vision was derived from their shared worldview and resulted in a clear distortion of reality. Although this is an extreme example, it does raise the question of whether other worldviews, no matter how firmly held by groups of people, provide an accurate or distorted view of the world.

Lest you think that this reasoning only pertains to poems or cult groups, let's look at an aspect of a worldview that characterized mainstream American culture for over a century and still can be found in segments of the culture. Specifically, consider the shared view regarding African Americans that many white Americans held or still hold. The shared view, or stated differently, the shared schematic representation of the inherent inferiority of this group, led to a clear distortion of how African American individuals were seen (or should I say unseen). The opening paragraphs of Ralph Ellison's classic novel *Invisible Man* provides a poignant example of the impact that this unseeing can have on an African American.

> I am an invisible man. No, I am not a spook like those who haunted Edgar Allan Poe; nor am I one of your Hollywood-movie ectoplasms. I am a man of substance, of flesh and bone, fiber and liquids—and I might even be said to possess a mind. I am invisible, understand, simply because people refuse to see me. Like the bodiless heads you see sometimes in circus sideshows, it is as though I have been surrounded by mirrors of hard, distorting glass. When they approach me they see only my surroundings, themselves, or figments of their imagination—indeed, everything and anything except me.
>
> Nor is my invisibility exactly a matter of a bio-chemical accident to my epidermis. That invisibility to which I refer occurs because of a peculiar disposition of the eyes of those with whom I come in contact. A matter of the construction of their *inner* eyes, those eyes with which they look through their physical eyes upon reality. (Ellison [1952] 1995, p. 3)

The inner eye, or cultural lens, is what we're talking about here. Clear distortions of reality can be found both in fringe groups and in mainstream cultural worldviews. It is far too easy to say that the worldview of the Seekers was distorted simply because they were some lunatic cult. Distortions can be found in many, if not all, cultural worldviews, including our own.

WHY CRITIQUE THE UNITED STATES' WORLDVIEW?

My main aim in this chapter is to articulate and evaluate the worldview of the United States. Why an emphasis on this particular worldview? Primarily because a major export of the United States, besides cars, candy, video games, smartphones, and clothing, is an American lifestyle based on a particular worldview. If this worldview is the root cause of climate change, then it is critical to know what this view is, see how it leads to the climate change problem, and take steps to change it. Ideally, if we can see more clearly and adopt a cultural lens that promotes harmony between human and environmental systems, perhaps other nations will follow suit. This has the potential to create a cumulative effect that reaches far beyond our borders.

When evaluating this worldview, the main question to consider is whether it helps us see better or whether it distorts our vision. If it is distorting our vision, we need to try to understand whether our "clear" vision is leading us over a precipice to our downfall. Does this worldview promote our survival? Is it linked with climate change? Might it lead to societal collapse? Moreover, other questions concern whether it promotes our well-being and happiness. We need to study this lens and acknowledge the impact it has on the way we see the world, the way we treat the natural world, and the way we react to environmental threats associated with climate change. We need to articulate the impact this worldview has on our lives and, on the basis of sound information, decide whether to accept or reject it.

In discussing the impact this worldview has on climate change, I employ the general model presented in chapter 1, focusing on how this worldview—which is typically characterized as one where people are disconnected from nature, self-absorbed, and concerned with self-enhancement—leads people to feel indifferent to nature (figure 2.1). This worldview also leads people to think that self-enhancement is best achieved through the acquisition of material goods, a thought process that results in individuals making a major contribution to climate change and being in disharmony with environmental systems.

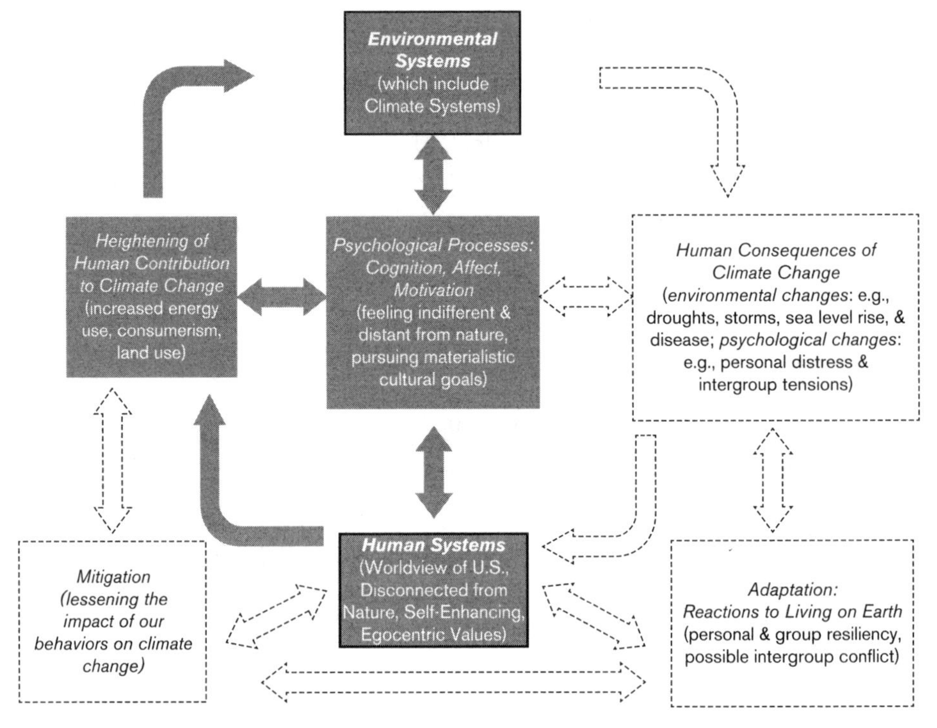

Fig 2.1. Adapted from Swim et al., "Psychology's Contributions to Understanding and Addressing Global Climate Change," *American Psychologist* 4 (2011): 242, fig. 1.

THE GENERAL ARGUMENT

There are two key points to keep in mind when considering how this worldview influences our sight. First, consider how a cultural lens that leads people to see separation between things can result in people *feeling indifferent* about harm directed at the natural world. Second, consider *how indifference in combination with specific kinds of cultural striving* can lead to climate change. These two key points, acting in unison, are leading us toward a climate tragedy.

Feelings of Separation, Distance, and Indifference

As a general rule, connectedness promotes compassion, empathy, and perspective-taking. Research in psychology clearly links these factors with caring and helping both others and the natural world (see Mayer and Frantz 2004, for a discussion). In contrast, feeling separate or disconnected from someone or something promotes indifference and lack of concern. When people feel connected to someone or something, the caring that accompanies this feeling typically leads them to feel inhibited from engaging in harmful behavior; whereas when people feel disconnected, they can engage in harmful behavior with psychological impunity.

This psychological sense of feeling connected or disconnected from someone or something is closely tied to the issue of how we can *distance* ourselves from others. Intellectually, physical and psychological distancing is a fascinating topic. In reality, tragedy is written all over it. When considering when humans are likely to harm another person, psychological and physical distancing is a critical issue. People generally are unlikely to harm those they feel close to or connected to. However, this reluctance to harm others evaporates when physical and psychological distancing occurs. The sense of "us" is replaced by a conception of "me" and this "other," and the likelihood of a person harming this "other" is far greater.

The argument I'm making is that humans do not engage in harmful behavior because they are inherently unfeeling killers or callous individuals. Contrary to what many people believe, a large body of evidence strongly argues that humans are not inherently aggressive, warlike, or bent on destruction (Fry 2007; Horgan 2012). For instance, in his book *Men against Fire*, S. L. A. Marshall (1947) discusses the results of interviews with hundreds of infantry soldiers who fought in the central Pacific and European theaters of war during World War II. Surprisingly, these interviews revealed that the soldiers were not dead set on killing the enemy. In fact, on average only around 15 percent of them actually fired at the enemy with intent to kill. Other research with soldiers who had suffered from psychiatric combat fatigue indicates that the main fear of these soldiers was not being killed, but rather killing.

Psychological research has come to a similar conclusion. Robert Baron (1977), after reviewing over three hundred studies in psychology on aggression, states, "Contrary to the views espoused by Freud, Lorenz, Ardrey, and others, aggression is not essentially innate. Rather, it seems to be a learned form of social behavior, acquired in the same manner as other types of activity and influenced by many of the same social, situational, and environmental factors" (p. 269). Distancing is one of these social, situational, and environmental factors.

In a book titled *On Killing*, Dave Grossman (2009) outlines four ways that we can distance ourselves from others: cultural distancing, moral distancing, social distancing, and mechanical distancing. *Cultural distance* refers to thinking of the "other" as an inferior life-form, which makes that person easier to kill. In this circumstance, the natural inhibitions that restrain people from harming others are removed. Wartime propaganda posters that focus on racial or ethnic differences often represent the "other" as a monster or some violent beast that needs to be destroyed. Grossman states, "The greatest master of this in recent times may have been Adolf Hitler, with his myth of the Aryan race: the *Ubermensch*, whose duty was to cleanse the world of the *Untermensch*" (p. 161), which posits a responsibility of superior humans to eliminate inferior, or subhuman, people.

Moral distance is evident when one group feels that their cause or purpose is superior to that of another group, or when they view the other group as having transgressed in some manner. Grossman put it in these terms: "Their cause is Holy, so how can they sin?" (p. 164). The legality of one's actions also plays into this. "The process of asserting the legitimacy of your cause" is a factor justifying an attack and enabling people to strike another (p. 166).

Social distance refers to seeing others as less than human on the basis of social class differences. The fact that some people can view others as "white trash" or "scum," solely on the basis of their perceived social standing within a stratified society, captures the essence of this form of distancing. Lastly, *mechanical distance* refers to the physical distance we place between ourselves and the harm done to another. Dropping a bomb from thirty thousand feet, launching a cruise missile from a battleship, and directing a drone strike from thousands of miles away are all examples of how physically removing ourselves from the pain inflicted on others makes it easier to harm them.

The worldview of the United States distorts our sight by distancing us from nature. It leads people to see themselves as separate and independent from one another and from nature. It distances people from nature by leading them to see themselves as culturally, socially, and morally superior to nature in a variety of ways. Moreover, the urban lifestyle that goes hand in hand with this worldview physically removes people from nature (Hofstede 1980). This distancing is one crucial element that leads people to feel indifferent about harm directed at the natural world. Thus, this first key point about distancing addresses the fundamental human/nature split that others have identified as the root cause of environmental problems (White 1967). From my perspective, this fundamental split results from a particular worldview that leads people to erroneously see themselves as dissociated from the natural world.

Cultural Striving for Superiority and Self-Enhancement

The second key point to keep in mind is that distancing alone cannot account for climate change. Being disconnected from nature simply makes it easier to engage in harmful behavior that results in climate change. For instance, consider what it might have been like to be a pharaoh in ancient Egypt, following the cultural script dictating that monuments, such as pyramids, had to be built. The goal was to build the pyramid. Part of the cost was the lives of slaves. In this example, the aim or goal of the pharaoh was not to kill slaves but to build a monument. The slaves were simply instruments used in the process of achieving this goal. Feeling disconnected and psychologically distanced from the suffering of the slaves, however, enabled the pharaoh to harm them with psychological impunity as he went about the task of building.

In our time, the cultural script is different. We strive for superiority and self-enhancement by trying to be more successful than others and acquiring greater material wealth. This striving comes at the expense not of slaves (although exploited workers do exist) but of the natural world. Thus, the second key point to keep in mind is that distancing in combination with cultural aspirations associated with a cultural script can lead to harm.

Stanley Milgram's (1975) psychological research on obedience to authority illustrates this point, highlighting the idea that the interplay between distancing and following the cultural script to obey orders can lead individuals to harm an innocent "other." Milgram initiated this research in order to understand how the atrocities committed by the Germans during World War II occurred. Instead of trying to explain those acts in terms of inherent human aggressiveness or a personality characteristic, such as the fascist personality structure associated with the Germans, Milgram wondered whether the fact that we are socialized to obey authority figures was responsible. He viewed this as a universal cultural script. In other words, Milgram explained what happened in Germany not in terms of "bad seeds" (i.e., something about the German people) but in terms of the "soil" we are all nurtured in. A disconcerting implication of his analysis is that what happened in Germany could happen anywhere if certain circumstances are present.

In one set of Milgram's now classic studies, individual volunteers were led into a laboratory and seated with another participant, who, unbeknownst to them, was actually a confederate of the experimenter. After introducing the study as examining the impact of punishment on learning, the experimenter let the two participants "randomly" draw lots to determine which one would be the learner and which the teacher. Actually, the drawing was fixed, and the naive participant was always cast in the role of the teacher, and the confederate always in the role of the learner. In a nearby cubicle, in what was a highly realistic setting, electrodes were then attached to the "learner." While the naive participant watched these electrodes being attached, the learner informed the experimenter that he suffered from heart problems. The experimenter explained to the learner that while the shocks could be painful, they should not cause any permanent damage.

Following this exchange, the "teacher" was seated in front of a shock generator. The experimenter presented him with a set of questions to ask the learner and informed him that he should shock the learner after each incorrectly answered question. The shocks started at 15 volts and increased in 15-volt increments, ending at 450 volts. Around the 300-volt level, the unseen learner yelled, apparently involuntarily, after the administration of each shock, complained of heart problems, and refused to answer any additional questions. When the teacher questioned the experimenter about the learner's refusal to answer the questions, the experimenter calmly informed the teacher that the lack of an answer was considered a wrong answer, and that the teacher should continue to ask questions and administer shocks. After the 330-volt level, an ominous silence occurred. There were no longer any involuntary screams from the learner. There were no complaints. It appeared that the learner was no longer conscious. When questioned at this point, the experimenter would inform the teacher that "the experiment must continue."

Before the study was conducted, experts were asked to predict the percentage of participants who would deliver 450 volts to an innocent learner. They predicted a tenth of 1 percent would do so, arguing that this would account for the number of sociopaths (i.e., "bad seeds") in the study. Actually, in many variations of the study, over 60 percent went on to deliver the maximum amount of shock, illustrating that individuals' choices related more to the circumstances present (i.e., "the soil") than to the personality of the participants (i.e., "the seeds").

One interesting variation on that set of studies illustrates the impact of distancing. In these studies Milgram examined participants' obedience as a function of how near or distant the "other" was from them. That is, in some settings the learner was behind a wall, physically removed from the teacher, as described above, while in other circumstances the learner was in the same room with the teacher. Milgram found that greater physical distance between the participant and the victim increased the participant's willingness to obey the experimenter's orders. Thus, simply ordering someone to do harm did not lead him to obey, nor did simply distancing him from the victim lead the participant to harm this innocent other. Rather, it was the combination of the participant's sense that he was required to obey the cultural script—that is, obey a legitimate authority figure—*and* the distancing of the participant from the "other" that maximized the harmful acts.

Keep in mind that a cultural lens, besides leading to a representation of the world, is associated with cultural striving. A worldview leads people to value certain things and not others, to strive to take certain courses of action and not others. Cultural worldviews involve a script. When considering a culturally valued course of action, people might not engage in it if the cost is too high. For instance, they might not trample friends in order to make it to the "top." Feeling strongly connected to these others would inhibit them from doing so. But what if their cultural lens led them to dehumanize these "others"? What if they no longer saw themselves as members of the same community, but instead saw these others as dirt? Similarly, what if their cultural lens led them to no longer see themselves as connected to or in community with the natural world? What if it led them to devalue it and to view trampling nature as nothing more than trampling dirt? They would feel indifferent about this course of action. Consequently, under these circumstances they might very well trample nature with psychological impunity as they strive to make it to a culturally defined "top," even though their collective actions may wreak havoc on the natural world in the form of the unintended consequence of climate change. Keep these thoughts in mind as we move on to a more detailed discussion of the cultural lens of the United States.

ARTICULATING AND CRITIQUING THE WORLDVIEW OF THE UNITED STATES

The United States' cultural lens is composed of many different elements. These elements, or beliefs, are interwoven and provide people with understanding, meaning, and direction in their lives. Rather than cover every element that makes up this worldview, let's consider five main aspects: the autonomous self associated with the individualistic worldview, a particular type of individualistic worldview that emphasizes hierarchy and superiority, our conception of progress, analytic as opposed to holistic thinking, and our belief in a just world.

These beliefs form the bedrock of our experience. The impact they have on our personal lives and collective life is often unseen. They operate on a subconscious level. Nonetheless, they are foundational to our cognitive structure, how we view our world, and what we strive for in this world. It's important to make what is implicit in our lives explicit, delineating in detail the nature of these beliefs and how they influence our sight and actions.

The Individualistic Cultural Lens: The Autonomous Self and Distancing

The first aspect of the United States' cultural lens that leads to distancing is the autonomous self that characterizes this individualistic worldview (see figure 2.2). This aspect of our cultural lens leads us to see self and other as discrete objects, disconnected from one another and from the natural world. Many researchers have emphasized that people in individualistic cultures are expected, and nurtured, to be independent and autonomous beings—that is, free entities who should not be constrained by others (Matsumoto 2001). Along with a focus on this free, independent self, there is a corresponding emphasis on the "I" and "me," and not on the collective "we" and "us."

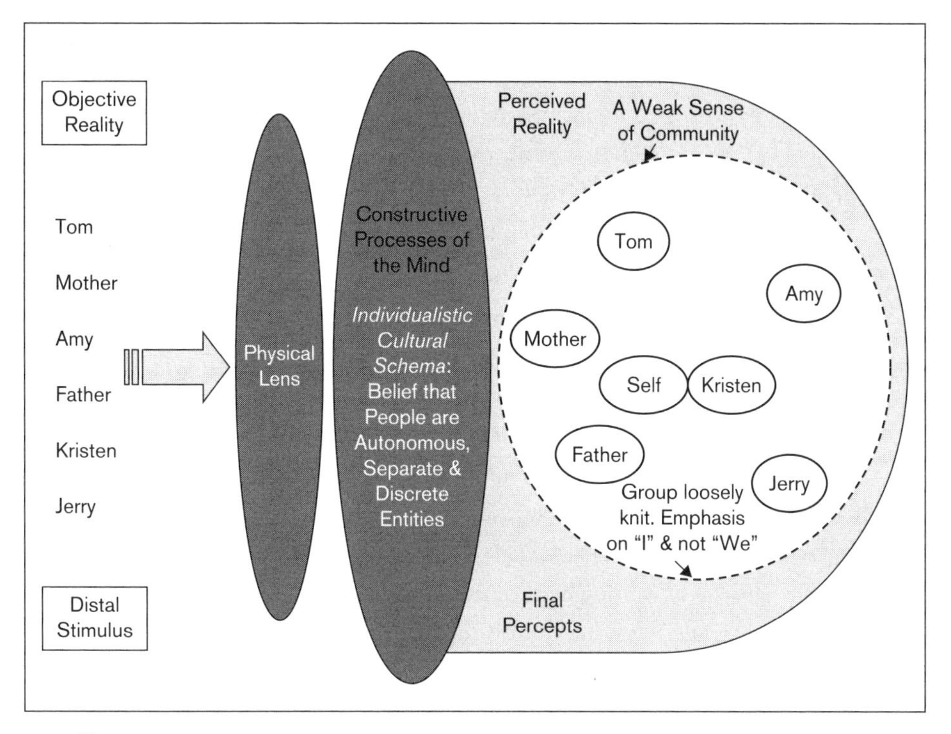

Fig 2.2

One's sense of self is thought to end at one's skin. Like other objects in the world, such as cups, cars, and chairs, a person's "object self" is thought to have clear boundaries. When you focus on one of these things, it is sharply contrasted with the background. It clearly stands out as the figure of your attention. In contrast, then, to the poet John Donne's assertion that "no man is an island," in this individualistic worldview we are nurtured and expected to be islands, clearly bounded and separated by narrow channels or wide seas from other individual islands (i.e., separate and more or less distant from strangers, acquaintances, friends, and family).

This separate, distinct, disconnected object self is often described as "decontextualized." (Oyserman and Lee 2010). That is, this free entity is unencumbered by ties to others. "No strings attached" is a good way of thinking of this orientation. People are expected to do what they want to do when they want to do it, and to not sacrifice their own personal interests for the interests of the group. So, if the normative expectations of the group differ from their own personal wants and desires, they are encouraged to follow their own inner compass and not succumb to group pressure. The conformist is frowned upon, viewed as someone without a backbone, and members of the individualistic culture are taught to stand up for themselves and not to lean on others.

In relating this discussion to distancing and the fundamental human/nature split, we may notice that this emphasis on the object self separates the individual not only from others but also from nature. The emphasis on "I" and not "us" can be extended to characterize an individual not feeling any connection or relatedness to the natural world. The image of the self as an island actually might be better represented by the image of astronauts in space, acting in a natural vacuum. Or, to use a Shakespearian image that highlights the more anthropocentric aspect of this individualistic orientation, we are actors on a human-made stage, where nature isn't even present. This is where the separation between self and "others" and between self and nature begins. Furthermore, examining this dissociation between self and nature, we begin to understand how compassion, empathy, perspective-taking, and caring for nature can diminish, and how indifference and harm to nature can become more probable.

The individualistic worldview is strongly associated with industrialization. For instance, in a famous cross-cultural study conducted by Gert Hofstede (1980), which included 117,000 participants from forty countries, he reported an astoundingly strong correlation of .8 between individualistic cultural worldviews and industrialization. Researchers have also observed that industrialization is strongly associated with a lifestyle spent in urban rather than rural settings and more time spent indoors than outdoors. For instance, in our industrialized country, an estimated 82 percent of Americans live in cities or suburbs (United Nations 2012), and Americans spend approximately 90 percent of their time indoors (Evans and McCoy 1998). This physical distancing of ourselves from nature is a form of mechanical distancing. By living in cities, we are not exposed to the harm being done to nature. Being physically removed is another way in which we can become indifferent about this harm.

Briefly critiquing the accuracy of this aspect of the United States' cultural lens, we may say that one of the basic tenets of environmental science is that we are all inextricably embedded within the biosphere in which we live (Coulson, Whitfield, and Preston 2003). There is no separateness between humans and the environments we inhabit. To think of a "decontextualized" self is a delusion. With every action we take, we are simultaneously affecting and being affected by the environment around us. "No strings attached" can never be, for from an environmental-science perspective we are woven into the fabric of the ecosystems we inhabit.

Similarly, psychological research emphasizes that we do not operate in a vacuum (Kenrick, Neuberg, and Cialdini 2015). A famous theoretical statement by Kurt Lewin—a social psychologist whose work has had a tremendous impact on modern-day psychology—that has heavily influenced theory development in psychology states that behavior, B, is a function of the person, P, in a perceived environment, E [i.e., $B = f(P, E)$]. This general approach to understanding behavior highlights our inherent embeddedness in the world around us. The emphasis on the interaction between the person and the environment underscores that behavior is always a "strings attached" type of phenomena. From this perspective, thinking of behavior solely as a function of the autonomous person, $B = f(P)$—as the individualistic worldview would have us see it—is clearly a misperception. In contrast to the individualistic worldview that highlights a decontextualized view of self, this interactionist approach argues that we are always influencing and being influenced by the context in which we are situated. Hundreds of studies in psychology illustrate this approach.

For instance, our worldview, which leads us to see ourselves as independent, nonetheless results from the influence other cultural members have had on us (Heine 2015). Worldviews are transmitted across time from one generation to the next. Adults nurture children to adopt these beliefs. Through this socialization process we learn and internalize accepted views of our culture, such as the view of our independence and autonomy. Thus, to think we are independent of others is as misguided as thinking that we do not depend on oxygen for life. After all, as children grow up within a culture, they depend on others to satisfy their basic needs, which include food, water, contact, and love. They also depend on others for a cognitive framework to make sense of their experience. All of these things highlight *not* their autonomy but their *relationship with* and *interdependence on* others.

Additionally, research in environmental psychology, which I discuss in more detail in chapter 4, illustrates how being in nature profoundly influences our thoughts, feelings, and actions. Our health, problem-solving ability, feelings of well-being, and even our sense of leading a meaningful life are all enhanced by being in nature. Moreover, our recovery from illness and stress has been linked to exposure to nature (Frumkin 2001). My point is that our settings influence us. Whether we spend more time inside or outside, in nature or in front of a computer, in an urban or a rural setting, our thoughts, feelings, and actions are not independent of the context in which we act. In summary, the individualistic worldview presents a clear distortion of reality, for it denies that we are inextricably related to the world around us as we affect and are affected by that world.

Lastly, it is profoundly interesting to consider how the characteristic of individualism, which encourages us to be separate from, not in relationship with, one another, may undermine our most basic social need for belonging (Fiske 2012). That is, to be connected to others is thought to be our primary aim in our social life. To experience reciprocal positive interactions with others is considered critical to our well-being. Thus, the individualistic characteristic encourages each of us to be an island when, in fact, our well-being is most strongly associated with meaningful relationships. This brings up the question of the extent to which this worldview undermines the quality of our lives.

Dion and Dion (1993), commenting on Western images of love and marriage, discuss the difficulties of developing intimacy when the surrounding culture values autonomy and devalues any form of dependency. In their research, they found "that 'self-contained individualism' was negatively related to reported caring, need, and trust of one's partner. 'Self-contained' individualists were also less prone to describe their experience of romantic love as rewarding, deep, and tender. They were more likely to view 'love as a game,' as a test of their skills and power in a love relationship"(p. 64). In other words, psychological individualism is related to emotional detachment. This is not a recipe for happiness.

Feelings of Superiority: Hierarchy and Distancing

To separate oneself from another in terms of "I"/"me" versus "you" or "it" is one step we take in distancing ourselves from another and from nature.

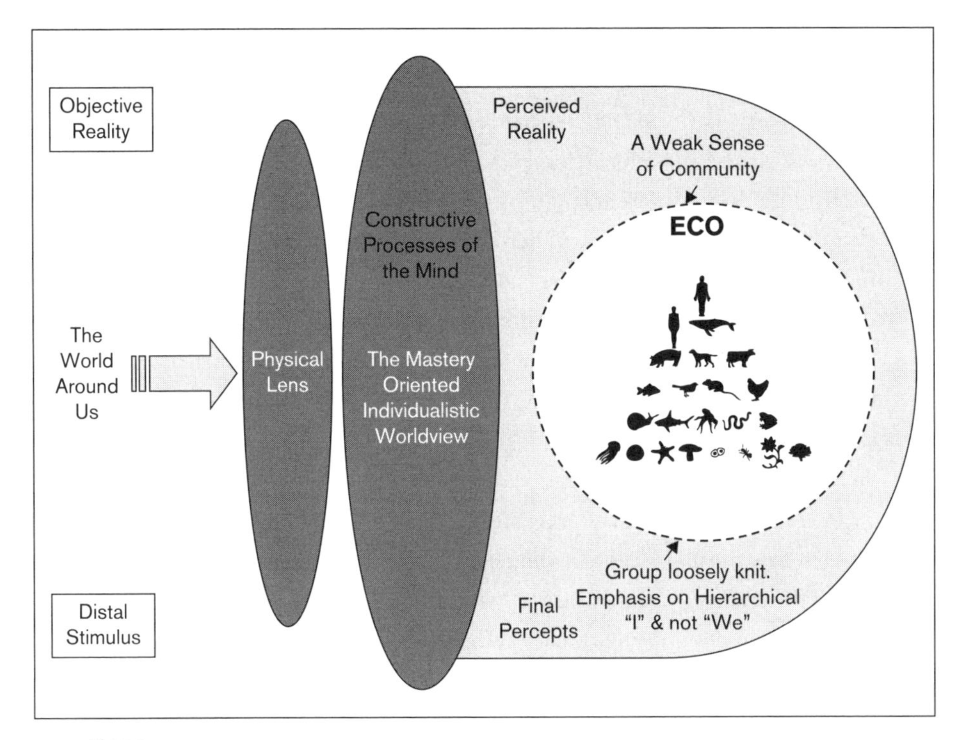

Fig 2.3

Another step takes place when the "I"/"me" is viewed as superior to the "you" or the "it." This is the egoistic characteristic of the mastery-oriented individualistic worldview found in the United States (see figure 2.3). As you read this section, you'll see that the view that humans are separate from *and* superior to nature only increases the human/nature split. It is critical to realize that the individualism mind-set comes in several different forms, and that some forms lead to greater distancing and disconnection from nature than others. In particular, the United States' cultural lens is composed of a particular type of individualistic worldview, a *mastery orientation*, that is especially pernicious in widening this split.

The evidence for this is nicely illustrated in research conducted by Shalom Schwartz (1994). Schwartz, an Israeli psychologist engaging in cross-cultural work, has conducted numerous studies on how cultural worldviews differ from one another (to name a few: Schwartz 1992, 1996, 2008). His first step was to create an exhaustive list of values. After extensively examining prior research, literature from around the world, and other sources, he was able to identify fifty-six values that serve as guiding principles in individuals' lives. He then developed a survey and, with the aid of colleagues, translated it into thirty different languages. The survey's format allowed respondents to indicate the extent to which each value served as a guiding principle in their lives. Following this, he and his colleagues administered the survey to eighty-six samples of respondents from forty-one cultural groups and thirty-eight nations. Eighty percent of the samples ranged in size from 150 to 300 respondents. Samples came from every inhabited continent and were written in thirty different languages, by adherents of twelve religions, as well as atheists. The size and scope of this project is truly remarkable.

After obtaining the data, he then analyzed it. In particular, he was interested in how the fifty-six values tended to cluster together in different groups. Stated differently, some of the fifty-six values can be thought of as sharing something in common with certain other values, but not with all others. Moreover, this analysis revealed not only how the individual values could be grouped together but also how the clusters are related to one another. Some clusters are positively related to one another, meaning that scoring *higher* on one set of values is associated with scoring *higher* on a second set. Other clusters are negatively related to one another, meaning that scoring *higher* on one set of values is associated with scoring *lower* on a second set.

Figure 2.4 depicts the results from this analysis. Examples of individual values are represented in the smaller font. For the sake of simplicity, instead of including all fifty-six of these values, I've included only the values I focus on in this book. To highlight how certain values cluster together and are separate from other value clusters, Schwartz provides a general name for each cluster. The general names appear in the figure in the larger, italicized font. For example, he provides the name "Harmony" for the individual value cluster "protecting the environment," "unity with nature," and "world of beauty." As for the individual value cluster "successful," "independent," "ambitious," "daring," "choosing own goals," and "capable," he provides the name "Mastery." And he has identified seven constellations of values ("Mastery," "Hierarchy," "Embeddedness," "Harmony," "Egalitarian Commitment," "Intellectual Autonomy," and "Affective Autonomy"). In other

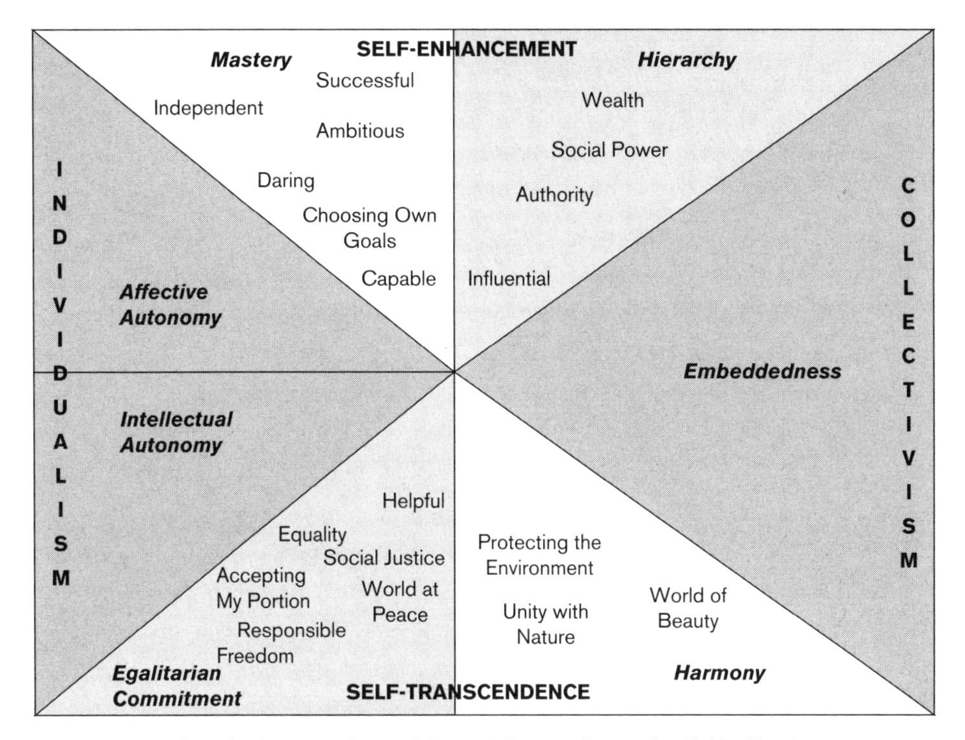

Fig 2.4. Adapted Schwartz Value Model. S. H. Schwartz, "Beyond Individualism/Collectivism: New Cultural Dimensions of Values," in *Individualism and Collectivism: Theory, Method, and Applications*, edited by U. Kim, H. C. Triandis, C. Kagitcibasi, S.-C. Choi, and G. Yoon (Thousand Oaks, CA: Sage, 1994), adapted from figure 7.1.

words, instead of thinking of fifty-six separate and independent value orientations that serve as guiding principles in individuals' lives, we can more generally think of seven different value constellations that guide humans on their journeys.

Given that these individual value clusters hold together empirically and act as guiding principles in individuals' lives, we can think of each of these value labels as a worldview. In this model, worldviews that are adjacent to one another are positively associated with one another. For instance, the worldviews of Mastery and Hierarchy are positively linked with one another. That is, a cultural worldview that endorses the individual values of "success" and "ambition" also tends to endorse the individual values of "wealth" and "social power." The cultural worldviews of Harmony and Egalitarian

Commitment are also positively associated with one another, meaning that a cultural worldview that endorses the individual values of "protecting the environment" and "unity with nature" also tends to endorse the individual values of "social justice" and "equality." Additionally, worldviews that are opposite to one another are negatively associated with one another. In other words, Mastery and Hierarchy are negatively associated with, or opposed to, the worldviews of Harmony and Egalitarian Commitment.

More generally, Schwartz simplifies this model even further by thinking of these clusters or worldviews as being aligned along two dimensions. One dimension is "Self-Enhancement / Self-Transcendence." For the purposes of our discussion, I've relabeled the second dimension as "Individualism/Collectivism" instead of what he refers to as "Autonomy/Conservatism." Given that our discussion primarily focuses on the former dimension, in figure 2.4 I've highlighted the worldviews associated with Self-Enhancement (Master and Hierarchy) and Self-Transcendence (Harmony and Egalitarian Commitment). It is important to note, however, that with reference to Individualism, in this figure you can see that Egalitarian Commitment is a form of Individualism that is more Self-Transcendent in nature, while Mastery is a form of Individualism that is more Self-Enhancing, or egoistic.

The greatest distancing between individuals occurs in the Self-Enhancing value orientations of Mastery and Hierarchy. On the subject of Mastery, the goal of the ambitious and success-driven individual is to achieve and surpass others. And in a related vein regarding Hierarchy, its neighbor, the goal of the ambitious and success-driven individual is to acquire greater wealth and power than others. This is very different from the cultural worldviews of Harmony, where unity with nature is emphasized, and of its neighbor Egalitarian Commitment, where people strive for equality.

This discussion of the Schwartz model is a prelude to the research that finds that the United States endorses a mastery-oriented individualistic worldview. In fact, regarding mastery, in a comparison of twenty capitalist countries, the United States was the second-highest-scoring country (Schwartz 2007). That is, members of the United States view the values of independence, ambition, success, daring, and personal capability as the most important guiding principles in their lives. The mastery worldview represents a self-enhancing, or egoistic, form of individualism. As for hierarchy, this same study found that the United States was the second-highest-scoring nation. These findings paint a picture of the United States as endorsing a worldview in which distancing is the norm. People aim to differentiate themselves from others, not just by seeing themselves as separate and discrete entities in the world, but also by being superior to others in terms of their accomplishments, success, wealth, and social power.

Underlining Schwartz's finding that self-enhancement is a pervasive tendency in the United States, hundreds of studies have demonstrated that Americans differentiate themselves from others and see themselves as superior to others (cf. Brown 1998; Sedikides and Strube 1997; Taylor and Brown 1988). These feelings of superiority come in many forms. American individuals believe they are not only more talented than others but also smarter, warmer, and more thoughtful (Alicke 1985; Brown 1986). They view themselves as fairer (Messick, Bloom, Boldizar, and Samuelson 1985) and even see themselves as better drivers than most other people (Svenson 1981). They perceive themselves as well above the fiftieth percentile in terms of exhibiting social graces and leadership abilities (Alicke 1985; Dunning, Meyerowitz, and Flolzberg 1989). Americans also self-enhance and inflate their sense of self-superiority by taking more credit for their successes than for their failures (Ross 1981). When assessing how much they contributed to a group's success, people often see themselves as contributing more to the success than the other group members (Savitsky, Boven, Epley, and Wight 2005). As Neff (2008, p. 96) points out, in America self-esteem plays a critical role, and in "American culture, at least, having high self-esteem means standing out in a crowd—being special and above average (Heine, Lehman, Markus, & Kitayama, 1999)."

This desire to feel special and superior extends to anything associated with a person's self. Americans view the groups they belong to as superior to other groups (Tajfel and Turner 1986), their friends and loved ones as better than most other people (Brown 1986; Murray and Holmes 1993, 1997), and their romantic relationships as more special than the romantic relationships of others (Van Lange and Rusbult 1995). This is a self-centered, egoistic orientation.

As you might imagine after reading the first part of this discussion on how values tend to be related to one another, the Schwartz (2007) study found that the United States scored lower on Harmony than any other nation and second-lowest on Egalitarian Commitment. That is, striving to get ahead and establish personal superiority to others is not conducive to striving to protect the environment or striving to feel unity with nature (Harmony). Striving for self-enhancement is also negatively associated with striving for social justice and social equality (Egalitarian Commitment). Box 2.1 provides a summary of the findings from Schwartz (2007).

How is this mastery-oriented individualistic worldview expressed in the United States? Individuals tend to be rank-ordered from first to last on a wide variety of dimensions, and those at the top receive the awards. For instance, the high school student with the highest grade-point average is singled out to give the valedictorian address at commencement. The top athletes are differentiated from the lesser players. If self is an island, in this type of culture you want your island to be the envy of all the others. Separating self from others (i.e., being a separate and independent self) means striving to be viewed as distinctly better than others in some way. In the Olympics of life governed by the mastery-oriented individualistic worldview, the goal is to get the gold and stand on the highest pedestal or platform at the awards ceremony. Thinking of it in this way, you can see that it is not simply individualism that leads to distancing of the individual from nature, but the combination of individualism and placing oneself on a platform raised far above the natural world.

Considering how feelings of superiority affect the fundamental human/nature split, cultural distancing (i.e., seeing oneself as a superior life-form) easily comes to mind. In science, we place ourselves at the top of the evolutionary ladder. In certain religious traditions in the United States, even associating humans with nature through evolutionary thought is blasphemy (Masci 2014). Moreover, philosophical and scientific traditions, which provide a mechanistic view of the natural world, reinforce the notion of human superiority to nature. Indeed, some traditions argue that while humans have life and free will, nature is inanimate and functions much like the gears of a watch. Given this sense of seeing ourselves as a superior life-form, is it any wonder that we can feel indifferent about harming nature?

BOX 2.1. SUMMARY OF FOUR VALUE ORIENTATIONS AND THE U.S. RANKING ON EACH. FROM S. H. SCHWARTZ, "CULTURAL AND INDIVIDUAL VALUE CORRELATES OF CAPITALISM: A COMPARATIVE ANALYSIS," *PSYCHOLOGICAL INQUIRY* 1 (2007): 52–57.

Mastery Value Orientation (United States ranking: *High*)

- Encourages groups and individuals to master, control, and change the social and natural environment through assertive action. Others are often viewed in terms of how they can be used for one's own purposes. Exploitation of resources in the interests of progress and change takes precedence over preserving natural resources and protecting the immediate welfare of people whose interests conflict with one's own.
- Cultures that rank high in mastery emphasize ambition, daring, and success.

Hierarchy Value Orientation (United States ranking: *High*)

- Relies on differential, hierarchical allocation of roles and resources to groups and individuals as the legitimate, desirable way to regulate interdependence. People are expected to meet role obligations, accepting external social control. People seek their own competitive advantage without concern for others. The unequal distribution of resources privileges the strong. Does not promote collaboration with others.
- Cultures that rank high in hierarchy emphasize wealth, social power, and authority.

Harmony Value Orientation (United States ranking: *Low*)

- Encourages groups and individuals to fit harmoniously into the natural world and social world, avoiding self-assertion aimed at exploiting or changing it.
- Cultures that rank high in harmony emphasize such values as experiencing the world at peace, unity with nature, and protecting the environment.

Egalitarian Value Orientation (United States ranking: *Low*)

- Emphasizes socializing individuals to accept others as morally equal, to transcend selfish interests, and to cooperate voluntarily with others out of an understanding of long-term mutual interests. It depends on the internalization of other-concern values.
- Cultures that rank high in egalitarianism emphasize such values as equality, social justice, accepting one's own portion, and freedom.

Moral distancing also comes into play here, for we place our rights and desires far above the rights of nature (if we even believe that nature has rights). We feel entitled to strive to surpass others, and if this involves the exploitation of nature, then so be it. This narcissistic sense of entitlement, which has reached epidemic levels in the United States (Twenge and Campbell 2009), has been empirically demonstrated to distance people from nature, especially when this aspect of narcissism is combined with a heightened sense of the object self (Frantz, Mayer, Norton, and Rock 2005).

Overall, when considering the fundamental split between individuals in the United States and nature, the picture that emerges is that of a people who embrace a specific form of an individualistic worldview that initially distances the self from nature by an emphasis on the discrete "I." The industrial/urban lifestyle associated with this worldview further physically (mechanically) distances people from nature. Lastly, this mastery-oriented individualistic worldview, with the associated sense of cultural and moral distancing, only adds to the human/nature split and enables people to actively harm nature with indifference and psychological impunity.

The Big Picture: Shifting from an Ego to an Eco Orientation

The first step, then, of differentiating the world into an "I" and "other," in combination with steps to inflate the "I" and diminish the "other," leads to an ever-increasing divide between how we see ourselves and how we see our relationship to nature. Nature becomes further and further removed from what some refer to as our "scope of justice," meaning that considerations of fairness and concern are not extended to the natural world.

In summary, the Schwartz model illustrates that being more concerned with establishing one's superiority to others leads to less concern for curbing climate change and, in that way, protecting the natural world. When one is more self-absorbed and intent on egoistic pursuits, the fate of the natural world gets crowded out of one's awareness and is not a guiding principle in one's life. Furthermore, think about how concerns for success and wealth undermine concerns for equality and social justice. In the Schwartz model, Egalitarian Commitment, with concern for equality and social justice, is located next to and is positively associated with Harmony; it is negatively related to Mastery and Hierarchy. It isn't a major stretch to consider how a mastery orientation not only banishes one's concern for nature but also banishes equality and social justice concerns, which include environmental rights and the rights of species to exist. It also isn't a far stretch to consider how crowding out issues of social justice leads people to be less concerned about the world they leave to their children. After all, inclusion of others or the natural world in one's scope of justice leads to concern and action (Opotow 2005); and with a self-enhancing, egoistic, more self-absorbed worldview orientation, the ethics of extending concern to others or to the natural world is lost.

If you think about this model and take it seriously, it illustrates how this worldview blinds people to caring for the natural world. More than that, however, it portrays a culture that is opposed to thinking of self as equal to or in unity with nature. This is a major problem. It may very well be the problem of our time. How do we shift from a mastery-oriented individualistic worldview to one where the values of harmony and egalitarian commitment are endorsed? Stated differently, how do you shift from an *ego* orientation, where self is hierarchically superior to other life-forms, to an *eco* orientation, where self is on an equal footing with, and in community with, nature? This is the shift that needs to occur. The land ethic worldview (discussed in chapter 3) represents such a worldview (see figure 2.5).

The superiority aspect of this individualistic worldview bears critiquing. Clearly, some people are better at some things than others. We are not all equal in our talents and abilities. Some people work harder than others, consistently putting more effort into their pursuits. Individual differences abound. But difference does not necessarily mean an inherent superiority of one person or group relative to another person or group. When people see a

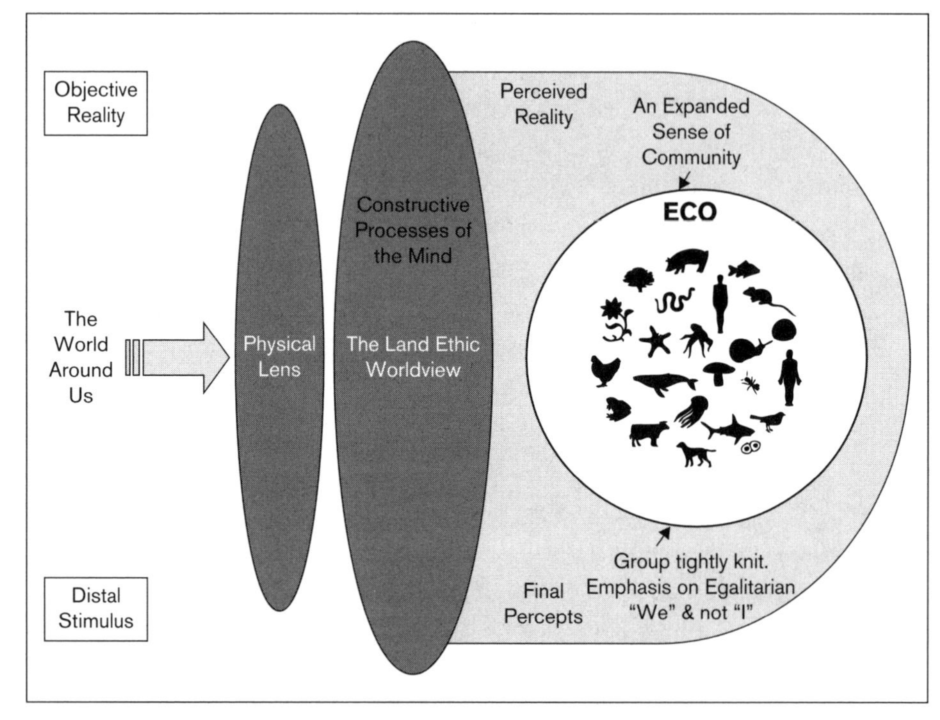

Fig 2.5. The land ethic worldview, an ecological worldview.

difference linked with self, and distort their view to inflate their sense of self-worth, differences that may not exist or that have no inherent meaning can take on an implied value that becomes a justification for a personal or social hierarchy that is unwarranted.

Look at the ego image again (refer back to figure 2.3). I can assure you that in the minds of many, the dark human figure at the top of the hierarchy in the ego image is not a Black man, or a woman for that matter. Examining that figure more closely, notice the other human figure, beneath the main one. Who is that figure? Gould's powerful book *The Mismeasure of Man* (1996) documents the fact that, historically, people have been categorized and placed within a hierarchical relationship to one another. And when you think of the damage done on the basis of racial or gendered feelings of superiority, think of the even greater distance placed between people—at the top of the ego pyramid—and all of nature, which lies beneath them. If the human divisions are great, the human/nature split is a chasm.

The issue of how the sense of superiority, or the self-enhancing orientation, relates to the basic drive to belong is also worth considering. Recall Dion and Dion's comments about Western images of love and marriage, and how individualism presents an obstacle to forming lasting, meaningful, intimate relationships. Add on the further distancing between individuals that is associated with the desire to feel superior to others, and it's not hard to see that the relationship problems become even further exacerbated.

Moreover, when considering personal development, psychodynamically oriented psychologists suggest that healthy development involves movement away from one's own self-interest, or egoistic concerns, toward greater concern for others (Adler 1956) and caring for the world (Erikson [1959] 1980). Similarly, the humanistic psychologist Abraham Maslow (1954) suggests that striving for self-transcendence, characterized by moving beyond a concern for only oneself and feeling a holistic, spiritual concern for others and nature, is a higher form of personal growth than are egoistic needs for self-esteem and recognition. Cognitive theorists of moral development (Kohlberg 1969; Gilligan 1982) also argue that more advanced stages of moral development involve moving away from self-centered concerns for reward, punishment, and social acceptance, and toward general concern for the well-being of others and of society as a whole.

Lastly, consider how people gain acceptance and the admiration of others through their ambition, success, and the concomitant expression of success in wealth and material goods. This self-enhancing orientation is associated with an ever-increasing materialistic consumption that is directly related to carbon emissions. Overall, while feeling superior to others may feed our egos, overinflating our sense of self-worth may lead to a self-indulgent lifestyle that distances and disconnects us from others and the natural world. Gaining possessions but losing connections can lead to behaviors harmful to others, ourselves, and the natural world.

The Idea of Progress: Striving for Personal Success and Material Wealth

Given the materialistic emphasis that characterizes the United States worldview, the view that things are getting better and better is generally how people in our culture think about progress. People expect that each generation will lead a better life than the previous generation. The march of progress involves better health, longer lives, and a higher overall standard of living. People expect enhanced purchasing power, which usually means the increased opportunity to buy things they expect will enhance the quality of their lives. This materialistic individualism lies at the heart of our culture of consumption. Striving for success is largely equated with wealth. Happiness and individual progress are largely associated with having larger homes, fancier cars, the latest technological gadgets, and other things which promise that, once we own them, we will be esteemed by others and will lead more fulfilling lives.

As David Myers (2000) states in his fascinating article "The Funds, Friends, and Faith of Happy People," the present-day American dream is succinctly summed up in the phrase "Life, liberty, and the *purchase* of happiness" (p. 58). People generally believe they need money to purchase things to be happy. This belief, of course, is based in part on the reality that we do need a certain amount of money to simply meet our basic needs. After all, a roof over our head, food on the table, clothes, and money for doctors and dentists all contribute to happiness. The question, however, centers on whether, once these needs are met, increasing amounts of money lead to ever-increasing levels of happiness. People certainly believe they do. When individuals are asked how much money they need to be happy, they typically answer "more." Generally, the answer is: "If I only made twice as much as I make now, I'd be happy." When late adolescents in the United States are asked how important financial success is to them, approximately 70 percent respond that it is either "very important" or an "essential" goal in their lives (Myers 2000). So the pursuit of ever-higher wages and more material wealth goes on and on.

It is also important to consider the symbolic nature of the things we buy. The idea that material goods help us lead more fulfilling lives needs to be unpacked. Certainly some of the things we buy make us more efficient at our tasks. By saving us time, these material objects allow us to lead more fulfilling lives, because we have more time to do the things we want to do. But when you think about the purchase of a home far larger than any family would actually need, or a car fancier than any practical purpose would require of it, or expensive apparel that may have nothing to do with practicality or efficiency, another issue comes up. Namely, we need to consider the symbolic role these material objects play in our lives. Owning the large home says something about the owner. Similarly, owning the latest version of some prestigious object lets an individual convey a message about herself: that she has succeeded, that she is a person of worth, not just in the monetary realm, but also in the social realm. Such things become the medals that mark our social status.

Let me back up for a moment and place this discussion in the context of the other characteristics that have already been covered. The main picture I have drawn illustrates how the characteristics of individualism, especially the mastery form, distance us from nature. Recall that distancing enables people to harm nature with psychological impunity. As the argument goes, if people distance themselves from nature, they won't feel inhibited from trampling it as they strive for a desired goal. This is where the characteristic of progress comes into play.

This cultural striving for more and more material things comes at a cost, and not just in terms of your wallet. This scrambling to the top harms the natural world. Heating and cooling homes far larger than necessary, lighting them, and consuming electricity to power our electronic devices, all directly contribute to CO_2 emissions. Similarly, the use of an automobile to drive forty miles to work or even one mile to the grocery store also produces CO_2 emissions.

Buying a new car also indirectly produces CO_2 emissions. In fact, there is embodied energy in everything we buy, meaning that CO_2 emissions result from the manufacturing, packaging, and delivery of products. In a throwaway society such as ours, which covets the latest version of any product, purchasing something time and again only magnifies these indirect contributions to CO_2 levels. Indirect contributions to CO_2 also come from eating a grapefruit in Cleveland that was grown in Florida. The transport of the commodity contributes to CO_2 emissions. Together, the direct and indirect contributions are staggering. In 2003, each household in America emitted, on average, an estimated fifty-nine tons, or 118,000 pounds, of CO_2, which was six times greater than the amount produced by the average household in the rest of the world (Hinkle Charitable Foundation 2006).

Like the pharaoh striving to build a monument, Americans striving for social status can, because of the human/nature split, harm the natural world with psychological impunity. We may not be trampling our friends to get to the top, but we are trampling nature as we strive to reach a culturally derived goal.

Research has shown that there is evidence for the cultural viewpoint that increased wealth is associated with increased happiness, up to a point. Comparisons of nations do show that people in wealthier nations tend to be happier than people living in poorer nations (Diener, Diener, and Diener 1995). When considering this relationship, however, it is important to think about how rich and poor nations vary in other ways as well, which could account for why people living in rich countries express greater happiness. For instance, rich nations tend to be more democratic than poorer nations, and people may feel happier in these nations not because of wealth but because of the political system under which they are living.

More telling is the research that looks at *within*-nation factors that contribute to individuals' happiness. For instance, consider figure 2.6, which examines the relationship between income and happiness (Myers 2000). As you can see, income has steadily increased in the United States since 1946. Notice how, in contrast, happiness has remained at the same level over this time period. This study strongly argues that these two factors are largely independent of one another. The conclusion from this study is that increases in income have no bearing on how happy people feel. This is not an isolated finding. Generally, research suggests that after people have enough money to take care of their basic needs, monetary gain does little to increase happiness (cf. Diener and Seligman 2004). Overall, then, the cultural viewpoint that wealth is associated with happiness is largely an illusion.

Tim Kasser has researched and commented extensively on the impact of materialistic striving on interpersonal well-being and on the negative implications of this striving for the natural world. Taking a close look at a review article that Kasser and colleagues (Kasser, Cohn, Kanner, and Ryan 2007) wrote on these issues, they state, "Studies show that materialistic values are associated with lower generosity (Kasser, 2005), as well as fewer prosocial (Sheldon and Kasser, 1995; McHoskey, 1999) and more anti-social activities such as cheating and petty theft (Kasser and Ryan, 1993; Cohen and Cohen, 1996; McHoskey, 1999). The importance placed on goals for financial success is also associated with greater disagreeableness (Roberts and Robins, 2000),

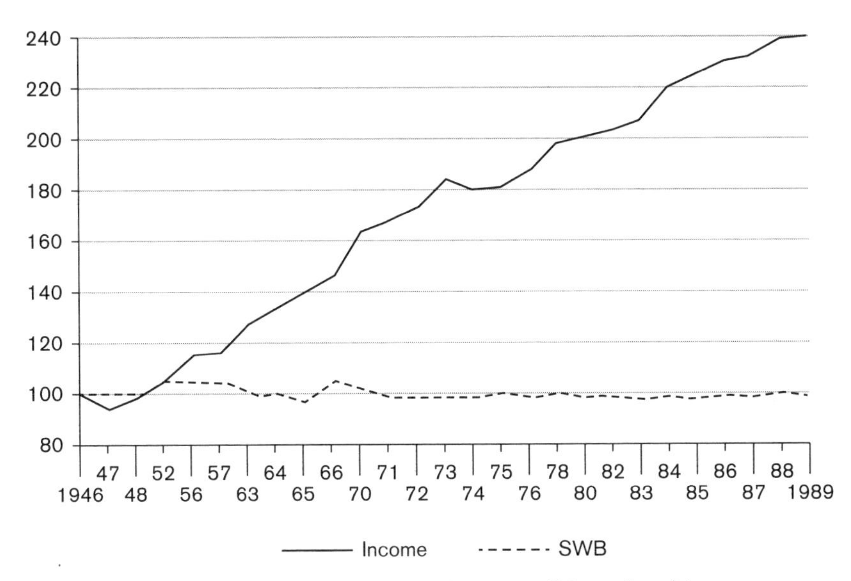

Fig 2.6. A representation of U.S. income and subjective well-being (SWB) from 1946 to 1989. Income is the percentage of after-tax disposable personal income in 1946 dollars (adjusted for inflation). Measures of subjective well-being are based on reports of happiness as a percentage of the 1946 values. Adapted from D. G. Myers, "The Funds, Friends, and Faith of Happy People," *American Psychologist* 55 (2000): 61, fig. 5.

lower empathy (Sheldon and Kasser, 1995), more Machiavellian tendencies (McHoskey, 1999), and more racial prejudice (Duriez, Vansteenkiste, Soenens, and DeWitte, 2005)" (p. 10). They also comment on how "financial success goals oppose those for community feeling, affiliation and self-acceptance. That is, concern for wealth and possessions conflicts with working 'to improve the world through activism or generativity,' 'having satisfying relationships with family and friends,' and feeling 'competent and autonomous'" (p. 8). Commenting on research by Lane, they state that "as Americans have pursued the aims of materialism and wealth, they have at the same time experienced 'a kind of famine of warm interpersonal relations, of easy-to-reach neighbors, of encircling, inclusive memberships, and of solid family life' (Lane, p. 9)."

Teresa Amabile's (1983, 1996) research also demonstrates that the pursuit of material rewards and social recognition undermines people's creativity. Instead of highlighting aspects of individuals' personality that can be linked to creativity, she examines the social contexts that people act within and the motivations that are encouraged within different contexts. In her work, she particularly focuses on the distinction between intrinsic and extrinsic motivation. Intrinsic motivation involves people being energized to engage in a task out of their personal curiosity, interest, and love of the task. In contrast, extrinsic motivation involves people being energized to engage in a task in order to receive a monetary reward or social recognition. She has found time and again that intrinsic motivation promotes creativity and extrinsic motivation undermines it. This research has led her and social commentators, like Alfie Kohn (1999) in his book *Punished by Rewards: The Trouble with Gold Stars, Incentive Plans, A's, Praise, and Other Bribes*, to question the wisdom of educational and business practices that rely so heavily on trying to increase performance by extrinsic factors. Thus, personal growth and well-being in the form of personal creative pursuits is another aspect of human functioning that is being sacrificed for materialistic gain.

As for the negative impact materialistic strivings have on the natural world, the review of the literature by Kasser, Cohn, Kanner, and Ryan (2007) points out how "people espousing more materialistic concerns express less love of the natural world (Saunders and Munro, 1999) and engage in fewer behaviors that benefit the environment (Richins and Dawson, 1992; Brown and Kasser, 2005; Kasser, 2005)" (p. 11). Additionally, materialistic concerns are associated with a larger ecological footprint (Brown and Kasser 2005). Moreover, in citing and commenting on Burroughs and Rindfleisch's (2002) work, Kasser, Cohn, Kanner, and Ryan comment on how the "evidence shows that increasing concern for wealth, power, and personal achievement corresponds with less concern for 'Understanding, appreciation, tolerance, and protection for the welfare of all people and for nature'" (p. 8).

The picture that emerges when critiquing the materialistic emphasis that characterizes the United States worldview is one where people are following an illusionary path, on which their striving for happiness, meaningful relationships, and personal growth and well-being leads to satisfaction at the initial steps, but which then disappears. Lured by this initial satisfaction associated with food on the table, clothes, a roof over their heads, and some form of transport, people continually strive for possessions symbolic of prestige, which only feed an egoistic, self-centered yearning for self-esteem and social status. And, as we've seen, this more egoistic or self-absorbed striving never truly fulfills the basic social need for belonging and the more mature needs for creativity and extending concern beyond one's self. Nonetheless, people travel on and on, led by the promise that is never fulfilled. The fact that people also tread over and destroy the natural world only adds to the tragedy of these misguided steps. Overall, what we've seen so far is that self-contained individualism, the striving for individual superiority, and the materialistic trappings that mark one's success and superiority all come at a high cost to one's self, others, and the natural world.

Analytic as Opposed to Holistic Thinking: Seeing Parts, Not Wholes

Another characteristic of individualism that corresponds to the United States' worldview is an emphasis on analytic as opposed to holistic thought, or systems thinking (Nisbett 2004). Analytic thinking focuses on parts. For instance, when asked what makes a car go, besides gas, a person might highlight engine parts like the carburetor, spark plugs, and cylinders. From this perspective, the self, like the engine, is simply another object whose workings can be broken down into smaller parts that make it go. When considering the self, however, the parts are talents and personality characteristics. For instance, when asked why an individual did something, it is not uncommon for people to explain the behavior in terms of the person's level of extraversion or shyness, or some talent or lack of talent the person has in a specific area.

This is not big-picture thinking. Relationships and the systems that operate on broader scales often go unnoticed. This reductionist, analytical form of thought may prevent people from seeing how they are in fact connected to other parts of a system, and how their actions affect broader systems such as the environment. Individuals' focus on the parts may lead them to experience a greater distance, or disconnect, from a natural system within which they are embedded. They may fail to see relationships between themselves and the natural world. This inability to see the whole or patterns is another factor that can distance or disconnect people from nature, which operates in terms of systems and interacting parts.

Lezak and Thibodeau (2016, p. 144) present holistic thinking as an aspect of systems thinking. Systems thinking involves "an emphasis on holism (as opposed to reductionism), an expanded conception of causality (i.e., an appreciation of the fact that a vast array of interacting variables are often responsible for specific outcomes in complex systems), and recognition that systems are constantly changing in predictable and unpredictable ways (Checkland, 2012; Espejo, 1994; Richmond, 1993; Sweeney & Sterman, 2007)." In a series of three studies contrasting systems thinkers with more analytically inclined thinkers, they found that systems thinkers were more likely to recognize the risks associated with climate change and to value ecosystems. Emphasizing the importance of this type of thinking, they conclude that "an emphasis on systems thinking may be seen as an attempt to promote a kind of 'wisdom' among the general public with regard to environmental decision making (Schwartz & Sharpe, 2010; Staudinger & Glick, 2011; Sternberg, 1990). In this context, systems thinkers recognize the possibility of unintended consequences in making consequential decisions and think more broadly about the array of potentially relevant antecedents and consequences of any given decision or behavior" (p. 151).

As for evaluating whether analytic thinking is a distortion of reality, the main point is that it clouds our vision by limiting our perspective. In other words, in a world composed of *both* parts and wholes, to see only the parts, however accurately we see them, distorts our view of the world, for the relationship between the parts goes unnoticed. The adage "Don't lose sight of the forest for the trees" captures the error associated with this emphasis on parts instead of the whole. Granted, some problems lend themselves to analytic thinking. However, not all problems do, climate change being one of them; and if you try to tackle a systems-oriented problem with analytic thought, the result will in all likelihood be failure. Consider the car mechanic confronted with an engine that is malfunctioning: if the mechanic focuses only on the independent parts of the engine, to the exclusion of the parts' interactions with one another, the mechanic may arrive at an inadequate solution. In psychology, the emphasis is on flexibility in problem-solving strategies. From this perspective, the goal is to be skilled in both analytic and holistic problem-solving strategies and to know when to use a given strategy.

The Belief in a Just World: Moral Distancing and Denial of Impropriety

"You got what was coming to you." How frequently have we heard that phrase? Often it has the ring of truth to it. *Often* isn't *always*, however; and whether right or wrong, the underlying meaning of the phrase is that justice was served. Or, what about the phrases "What goes around comes around" and "You reap what you sow"? These phrases express a similar sentiment and illustrate another feature of our cultural worldview: the belief in a just world.

Just-world belief refers to people's conviction that we live in a morally just and fair universe. This isn't something we necessarily consciously think about, although we've certainly heard someone get upset when this belief is violated and something unfair has occurred. Viewing the world through this lens leads people to believe that good things happen to good people and bad things to bad people. This lens helps people clarify what they see in the world. The world seems less capricious, reducing the uncertainty of why things occurred and adding stability and understanding to the experienced outcomes.

On the face of it, the belief in a just world seems innocuous. After all, what can be wrong with thinking that we live in a moral universe? Melvin Lerner (1980; Lerner and Montada 1998), a psychologist who started studying this belief in the late 1960s, vividly points out some the difficulties with this belief. He sees this belief as delusional and capable of distorting our sight, primarily because it can lead to judgments referred to as "blaming the victim." Consider this for a moment. If you affirm that we live in a morally just world, then events that occur must reflect that justice. So, if something bad happens to someone, then it follows that somehow that person deserved it.

What Lerner points out is that victims often do not deserve their fate. Parents who lose a child in a tragic accident do not deserve it; health-conscious women and men who die from cancer do not reap what they sowed; abuse, rape, and beatings do not occur because people did something wrong, somehow asked for it, or deserved the trauma they experienced. Blaming these victims for the trauma they endure only heightens the tragedy of their experiences.

While this belief can lead to delusional thoughts, it is often comforting (as long as you're not the victim). After all, if you view yourself and loved ones as good people, then you can feel assured that good things are in store for you and the good people in your life. The future looks bright. The just are rewarded; good people prosper. With this understanding of who is rewarded and who is punished in the world, and as long as you conceive of yourself as good and interpret your acts as good, you're basically in control of your future and can trust in the inherent fairness and benevolence of the world. Is it really that surprising that people want to enhance themselves and feel good about themselves, others they are attached to, and the groups to which they belong?

Ironically, this delusional belief system is also associated with positive mental health. People who view the world in this way report greater life satisfaction and well-being, and less depressive affect (see Dalbert 2001; Lipkus, Dalbert, and Siegler 1996; Ritter, Benson, and Snyder 1990). Seeing the world through "rose-tinted glasses" apparently has its benefits. When tragedy hits home, however, the delusion of fairness is shattered, and individuals experience the loss of what some researchers call an assumptive world (Kauffman 2002). Waking up to a view of the universe as neither benevolent nor fair can be difficult. The challenge of losing beliefs that one has based on assumptions made about the world can seem like a personal earthquake, where the bedrock beneath one's feet feels unsteady and uncertain. Nonetheless, leaving the comfort of a delusion for a clearer vision can be helpful, especially if a clearer vision enables the person to avoid tragedy or to extend compassion instead of blame to others (and perhaps even to himself).

In relation to climate change, belief in a just world can lead to moral distancing. People can easily see their cause or their cultural striving as just and warranted, even if this striving involves trampling nature. Belief in a just world can also prompt people to deny that bad things can happen to them in the form of the threats associated with climate change. As a friend once said to me after discussing these threats, "I can't believe that God would let this happen to us." In her view of the benevolent universe, a benevolent God is watching over us; and, trusting in God's love for us, she couldn't see how these threats could possibly result in tragedy. Whether one's benevolent universe is associated with the presence of a God or not, to think that bad things cannot happen to us because we are good is a belief that disconnects or distances us from these threats. They are so far removed from our thought patterns that we can't even see them. And with this disconnect and distancing comes inaction. As the singer Bobby McFerrin sang, "Don't worry, be happy."

FINAL THOUGHTS

That the worldview of the United States distorts our vision, and that this distorted vision leads us to blindly go about our business at the expense of the natural world, is disturbing. Because this worldview magnifies the human/

nature split and divorces us from nature, we no longer direct compassion, empathy, and care toward it. In fact, in many ways we can think of ourselves as operating in an anthropocentric world in which only our wants and desires matter. Or, should I say, we are engaged in a self-enhancing and egocentric quest largely framed by a cultural worldview that claims material wealth is the symbol of success? The research which has found that wealth generally does not lead to greater happiness should give us pause. The research which has found that our worldview in many other ways distorts our vision and can be viewed as the cause of climate change should not only give us pause but also create a deep fear in the pits of our stomachs. For if we extend this argument, it means that in a very short time we have to change our worldview if we hope to avoid the worst that climate change has to offer. What's your sense of how quickly a worldview can change?

But if the worldview is the cause, then changing it is what needs to be done. Adopting a worldview in harmony with nature needs to occur. Moreover, adopting a worldview that does not distort our vision, and that frames our actions in such a way that our striving promotes both the well-being of the natural world and our own personal well-being, is central. An endless treadmill of "If only I made twice as much as I make now, I'd be happy," when "twice as much" is always somewhere off in the future, is a treadmill to nowhere.

Seeing Clearly Means Seeing Connections

Our sight can limit us. It can limit not only what we see but also our aspirations and dreams. Our lives are impoverished by placing too great an emphasis on material wealth and insufficient emphasis on the wealth associated with close personal ties to others and to the natural world.

As Einstein (2015, p. 15) stated years ago, "We experience ourselves, our thoughts and feelings, as something separate from the rest—a kind of optical delusion of our consciousness. This delusion is a prison for us, restricting us to our personal desires, and to affection, for a few persons nearest to us. Our task must be to free ourselves from this prison by widening our circle of understanding and compassion, to embrace all living creatures in the whole of nature and its beauty." And as David Orr observed, "The effort to secure a decent human future, I think, must be built on the awareness of the connections that bind us to each other, to all life, and to all life to come. And, in time, that awareness will transform our politics, laws, economy, lifestyles, and philosophies" (2010, p. 323).

Looking Ahead

We need to transition to a worldview that is positively associated with harmony. We need to be less self-absorbed, less narcissistic, and less detached from one another and from the natural world. We need to adopt a perspective that is flexible enough to see the parts *and* the wholes of the systems in which we live. Like the people who lived during the transition from a geocentric to a heliocentric worldview, we need to bring our vision into line with what present-day science is telling us—in this case, about our inherent embeddedness in the natural world. We need to see how the world works, in terms of parts and in terms of parts belonging to larger systems. Lastly, given our psychological needs, such as our fundamental need to belong, we need to understand that in order for us to lead psychologically healthy lives, we need to nurture reciprocally positive connections and not separation.

In this book, I reiterate the theme of feeling connected time and again. At this point, however, it is important to see how the worldview of the United States distorts our sight. The lens we are encouraged to internalize as our mind's eye leads us to strive for separation and to see separation where connections exist. This worldview also places a greater emphasis on extrinsic, as opposed to intrinsic, motivation. As we've seen, this lens does not promote personal growth and creativity, close interpersonal bonds, or harmony between human systems and environmental systems.

In the next chapter the focus shifts from evaluating the worldview of the United States and showing how it can viewed as the root *cause* of climate change and lack of caring for the environment, to elaborating on how other psychological factors influence our willingness to care for the natural world and our willingness to confront climate change. Instead of focusing on what leads to climate change, the next chapter considers the psychological *obstacles* that prevent people from trying to correct this threat. I also offer steps that might be taken to increase this helping, or proenvironmental, behavior.

CHAPTER THREE

The Emergency of Climate Change

Why Are We Failing to Take Action?

Why has there not been a greater public outcry for action to confront the environmental threat underlying climate change? A variety of psychological barriers to helping come into play. In psychology, researchers in the area of prosocial behavior (i.e., the area addressing when people are likely to help other people) have grappled with the nature of these barriers for decades. Latane and Darley's (1970) model of helping has guided much of the research in this area. Their model sheds light on why people, faced with the threat of climate change, sometimes help but often fail to engage in proenvironmental behavior.

Psychologists started to intensively investigate helping behavior when, late one night in March 1964, a young woman's clear distress call went unheeded. That was the night Kitty Genovese was walking home to her apartment in Kew Gardens in Queens, New York. Arriving outside her apartment building, she was attacked. She was raped for more than half an hour and then murdered. Although she cried for help and, according to one account, thirty-eight people heard her cries, not one person lifted a finger to assist her. Following her murder, people were perplexed and horrified. How could this possibly happen? Today, the scientific community has issued a clear distress call regarding the threat of climate change. Many people are perplexed and horrified that a large number of Americans are failing to take action. Why do people often fail to help? What are the barriers to taking action?

INTRODUCING LATANE AND DARLEY'S MODEL OF PROSOCIAL BEHAVIOR

Latane and Darley present a five-stage model of helping (refer to box 3.1). Their model proposes that individuals move through each stage before helping. Briefly, stage 1 concerns people noticing an event. When people are unaware that something has occurred, they will not help. In stage 2, people, having noticed the event, need to interpret the event as an emergency, as something that warrants help, before they are willing to help. As you'll see, there is often uncertainty about whether an event is an emergency. Under such conditions, people often rely on others to help them clarify whether an emergency has occurred. Stage 3 concerns people feeling a sense of personal responsibility to take action. In order to take action, people need to feel connected to the emergency. This sense of connection can arise in two ways: (1) from their feeling that they somehow personally caused the problem, or (2) from a sense of "we," or interrelatedness, that arises from feeling connected to the other. Feeling a sense of "we" gives rise to empathy, compassion, and caring. Stage 4 involves a person trying to figure out how to help, while stage 5 involves the person actually having the ability to carry out an action plan. Even when all the other criteria are met, if people can't figure out what to do, or if they think they lack the requisite skills to do it, then no action will take place. From Latane and Darley's perspective, the path to helping is not a simple one. People take multiple steps, and face many obstacles along the way, when deciding what to do.

Additionally, people are thought to engage in an overriding cost/benefit analysis while progressing through these stages: they continuously calculate the positives and negatives associated with getting involved. If the cost is perceived as too high or the payoff too low, they may simply opt out and not help. Costs involve the possibility of being physically harmed, but also include financial costs, time involvement, stress, and possibly giving up enjoyable activities. Similarly, benefits are not limited to material gains but also include reducing the distress one might experience at witnessing the emergency, praise one might receive for having helped, positive feelings associated with knowing one has done the right thing, and possibly feeling closer to others.

So, why didn't Ms. Genovese receive help? Thirty-eight people witnessed her cries in the night. They clearly noticed that something was

BOX 3.1. SUMMARY OF LATANE AND DARLEY'S STAGE MODEL OF HELPING.

Stage 1. Noticing the Event: A person who is not aware of the event will not help.

Stage 2. Interpreting the Event as an Emergency: A person who sees the event, but who doesn't see that something is wrong, will not help.

Stage 3. Feeling Responsible to Help: A person may see the event and interpret it as an emergency, but unless she or he feels connected to the "other" (a sense of "we-ness" and not "I"/"it") or sees herself or himself as causally related to the emergency, that person will not help.

Stage 4. Forming an Idea of What to Do: A person who is at a loss for what to do will not help.

Stage 5. Ability to Actually Perform the Act: A person who knows what to do, but who doesn't feel able to perform the act, will not help.

The Overriding Cost/Benefit Analysis: If at any time a person sees getting involved as simply too costly (i.e., as disrupting his or her life or being personally harmful or generally negative) and/or not beneficial (i.e., not personally rewarding), that person will not help.

happening. However, when interpreting the situation, did they clearly see it as an emergency? Did they realize her life was being threatened? Additionally, did they feel a personal responsibility to help her? Latane and Darley's research highlights stages 2 and 3 of their model, which illustrate how the presence of other people influences helping.

For instance, one interesting thing about stage 2 concerns the uncertainty often present in helping situations. Did the thirty-eight witnesses view the event as a life-or-death situation, or as a lover's spat? In the minds of the bystanders, it might have been uncertain that this was an emergency. How might they have reduced their uncertainty? One way would have been for them to rely on others to help them clarify the situation. Were other people acting as if it was an emergency? A problem arises, however, when bystanders present in a situation don't want to potentially behave foolishly by overreacting. This can lead them to "play it cool" and remain calm. If this occurs, other bystanders looking around and trying to reduce their uncertainty by observing the reactions of others may interpret the event as a nonemergency and fail to take action.

Perhaps more importantly in the Genovese case, did their awareness of other bystanders lead them to feel less personally obliged to help? This is a key aspect of stage 3 in Latane and Darley's model. Namely, that *increasing* the number of bystanders present at an emergency can *decrease* the likelihood that someone will help. Called "diffusion of responsibility," this concept highlights the idea that with more people present, personal responsibility becomes diffused throughout the group, so that each person feels less personal responsibility to take action. Stated differently, Latane and Darley argue that Ms. Genovese would have been more likely to receive help had only one person heard her cries, rather than thirty-eight, for the one person would have felt complete responsibility to do something.

In one classic study illustrating stage 2, three participants, none of whom were confederates of the researchers, were in a room filling out forms when smoke started to appear from under an adjoining door (Latane and Darley 1969). According to reports and reenactments of this study, which are quite humorous, the participants diligently continue to fill out the forms, sneak looks at each other to see how the other participants are reacting, remain calm, and watch the room fill with smoke. They do nothing to address the possible emergency. This study illustrates how others can influence how we define our immediate reality. Given the uncertainty in the situation, their inaction and calm demeanors led them, using the information provided by the calm demeanors of those around then, to define the immediate situation as a nonemergency. As for being humorous, it is funny to see people act this way, until you consider the implication that no one helped. In contrast, when alone, participants typically reported the smoke in a timely manner.

Theoretically, Latane and Darley's reasoning and research findings make good sense. Theory and research in social psychology highlights how others, under conditions of uncertainty, influence the way we see our world. For instance, Festinger's (1954) social comparison theory argues that people have a basic need to hold correct opinions and attitudes, and that when confronted with uncertainty, people are motivated to initially try to reduce their uncertainty by comparing themselves against a physical reality. For instance, when dressing in the morning, if I am uncertain about whether I should wear a sweater, I can reduce this uncertainty and think that I've made a correct decision by looking at the physical reality of a temperature gauge outside my window. If the temperature is below forty degrees, I'll likely consider the decision to wear a sweater the correct one. The physical reality provided by the temperature gauge is sufficient to reduce my uncertainty and establish my correctness.

A problem arises, however, when there is no physical reality against which to compare one's ideas. When thinking of the correctness of many aspects of our social life, such as the correctness of the guiding principles in our life, there is no physical reality against which to compare our values, attitudes, and opinions to establish correctness. This is where it gets interesting. Under these conditions, social comparison theory argues, people will compare themselves against other people to reduce their uncertainty and establish correctness. Stated differently, under conditions of uncertainty, others often define reality for us. Thus, if people aren't acting like there's an emergency, other people will be inclined to *not see* the situation as an emergency and, consequently, fail to take action.

In another classic study illustrating stage 3, participants were placed in separate cubicles, where they communicated to one another over an intercom system (Darley and Latane 1968). One participant was actually a confederate of the experimenter. During one of his turns to speak, he suffered a purported epileptic seizure. The study was set up so that there was either one other participant present, three others present, or five others present. In other words, group size was varied, and the main question was whether the confederate would be more likely to receive help when fewer or more people were present. The researchers found that group size did influence helping, an illustration of the diffusion of responsibility effect. The confederate was more likely to receive help when there was only one other person present than when there were three or five others present.

THE PSYCHOLOGY BEHIND PROENVIRONMENTAL BEHAVIOR

Applying this model to the issue of why people fail to take action to address climate change, keep in mind the impact of distancing and disconnection from nature, not only as a *cause* of climate change, but also as a *barrier* that prevents people from providing aid. In particular, if a sense of "we" is considered necessary in order for people to help (stage 3), psychological disconnection (i.e., feeling separate) from nature can be viewed as a major barrier. As I argued in chapter 2 and will discuss in more detail in chapter 4, feeling a connection to and communality with nature may be the pivotal step we need to take to promote harmony between human and environmental systems.

This argument is illustrated in figure 3.1. The self-enhancing, egocentric, mastery-oriented individualistic worldview of the United States leads people to feel disconnected from nature, which serves as a major barrier to mitigation. The reason why it is a major barrier is because the psychological processes (cognition, affect, and motivation) deriving from feeling disconnected lead people to feel indifferent to the well-being of the natural world. Moreover, this feeling of indifference enables people to engage in acts that harm the natural world (e.g., pursuing success, social recognition, wealth, and power) with psychological impunity. Thus, this worldview is both a cause of climate change and a major barrier that blocks people from engaging in proenvironmental actions.

Although I state that feeling a connection to and communality with nature may be the pivotal step we need to take to promote harmony between human and environmental systems, this step needs to be placed within the context of a multitude of other factors that serve as barriers to mitigation. Eliminating the mastery-oriented individualistic worldview removes one major barrier. Replacing it with a worldview that promotes helping is a major step, but many other barriers to action still remain. Given the wide assortment of barriers, an interesting question to consider as you read this chapter and the rest of the book is whether a change in worldview, in which people develop a sense of connection to nature, may prompt them to overcome some of these other barriers as well.

Generally, Latane and Darley's model helps us more clearly understand these barriers to mitigation (see Frantz and Mayer 2009). Focusing on

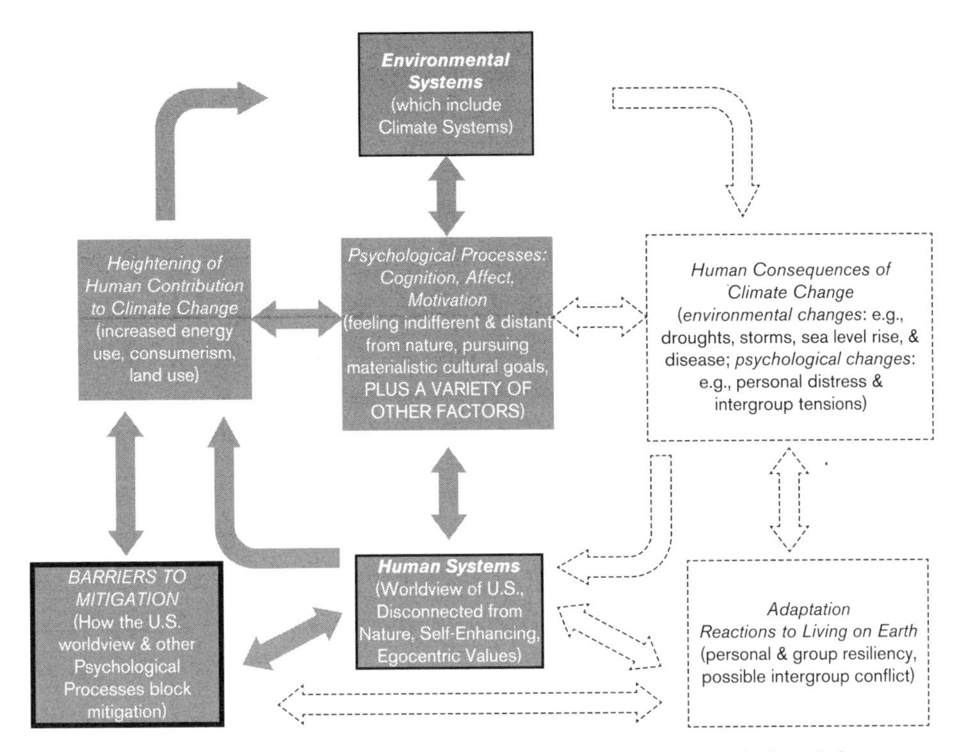

Fig 3.1. Model highlighting barriers to mitigation. Adapted from Swim et al., "Psychology's Contributions to Understanding and Addressing Global Climate Change," *American Psychologist* 4 (2011): 242, fig. 1.

barriers demonstrates that highlighting only factors that promote helping is limited, for factors that promote it always need to be considered in the context of those factors that block it. The beauty of the Latane and Darley model is that it provides a structure via which to consider this broad array of psychological processes that influence individuals' desire to address climate change (refer to Gifford 2011 for another interesting model).

Within this stage model, the interplay between the stages can also be addressed. Up to this point I've discussed the model only in terms of a linear progression from stage 1 to stage 5. That is, if the conditions for stage 1 are not met, then do not proceed to stage 2; if the conditions for stage 1 but not stage 2 are met, do not proceed to stage 3, and so on. The interplay, however, involves, for example, how feeling connected to nature (a factor affecting stage 3) may make people more likely to notice an event (stage 1) and/or more likely to interpret something they notice as an emergency (stage 2). When thinking of their model, it is important to not approach it in a rigid, sequential manner.

Illustrating this point, I always present a personal example to my social psychology class. When I lived in Los Angeles and was attending graduate school, I often relieved my stress by working out at a classic, 1920s-era L.A. hotel that had fallen on hard times. It still retained some of its old, Southern California charm. I could also get a cheap membership to use the workout facilities. It had a weight room, a whirlpool hot tub, and an outdoor pool. Typically, my workout routine involved an initial ten-minute soak in the hot tub, then swimming for half an hour, followed by another soak. One day when I began my routine, an old man was in the hot tub. After I returned from my swim, he was still in the tub, but now his head was almost underwater and his eyes were closed. I didn't know whether he had fallen asleep in the water or was simply relaxing and letting the jets massage his neck. In other words, I noticed him (stage 1), but was uncertain whether this was an emergency. Not wanting to overreact by grabbing him when he might only have been relaxing, I "played it cool" and tried to clarify whether it was an emergency situation by starting up an inane conversation. When he didn't answer, started snoring, and began to take in water, I knew it was an emergency (stage 2). Given that I was the only other person present, I felt completely obliged to do something (stage 3). I formed the idea of lifting him out of the water (stage 4), but he was too heavy. I tell my class that I'd made it to stage 4 (forming the idea) but was unable to perform the act (stage 5). I then look at the class and logically conclude that, because I was blocked at stage 5, I didn't help and left him to die in the hot tub.

This is where there is usually a stunned silence in the room and sometimes a gasp. My students have often looked at me in a bewildered manner at this point. But after I pause for a moment, I wave it off and tell them that what I actually did was think of another idea. That is, I returned to stage 4 (formed another idea). I went into the weight room and found two burly guys to help me help the old man. I emphasize to my students that thinking of the interplay of the stages—how we can move forward and back in the model—is important. Interestingly enough, when the paramedics arrived and the

BOX 3.2. STAGE 1. BARRIERS TO NOTICING THE EVENT

- Sensory limitations
- Our "ancient brain"
- Urban lifestyle / living indoors
- Technological devices
- Being self-absorbed
- Lack of place attachment
- Habituation
- Escapist activities
- General ignorance

club's staff members were huddled around the old man, who was now conscious and talking, I no longer felt responsible for him and faded into the background of the setting.

My point is to be flexible when thinking about these stages.

Stage 1. Noticing the Event

SENSORY LIMITATIONS It seems obvious that people will not help unless they first notice the event. But in relation to the issue of climate change resulting from increases from CO_2 levels, stage 1 takes on special meaning. After all, how can we expect people to react to something that they can't taste, see, or smell? This is a major issue. In the case of Ms. Genovese, her cries for help pierced the night. In other past emergencies, such as Pearl Harbor or 9/11, the emergency was a clarion call to action. In contrast, CO_2 has increased by approximately 95 parts per million since 1960. Graphing it, you can see a steady increase in CO_2 over time (see figure 3.2). The data, obtained from research sensors from around the world, could not be clearer. However, in our daily experience we're oblivious to these changes. Perceptually, we are literally blind to this steady increase. Given this, is it really that surprising that people have not rallied around the environmental flag?

Or, consider the temperature increase of .8 degrees Celsius that has occurred since 1880. Almost identical to the graph of CO_2 increases over time, this graph shows an unabated steady climb in temperature corresponding

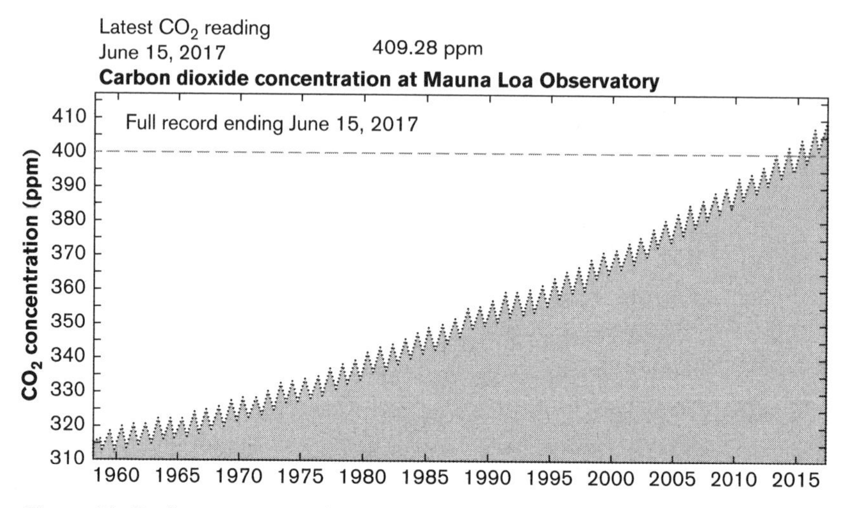

Fig 3.2. The Keeling Curve. From the Scripps CO_2 Program, Scripps Institute of Oceanography, scrippsco2.ucsd.edu.

very closely with industrialization (see figure 3.3). There is little ambiguity associated with this graph. But think about these changes for a moment. This is a .8 degree C change over 140 years. In other words, there has been a minuscule, fraction-of-a-degree change per decade over this period. As in the case of the CO_2 example, perceptually we are not equipped to notice this change. While the mechanical sensors that scientists have deployed around the globe can detect these changes, the senses we're equipped with as human beings are hard pressed to notice any change at all.

OUR ANCIENT BRAIN Besides the limitations of our senses, our ancient brain, which hasn't evolved substantially in the last five thousand years, and which still guides much of our behavior (Gifford 2011), presents another limitation. Primarily attuned to present circumstances, such as the immediate dangers around us, the immediate welfare of our tribe, and the availability of nearby resources (Ornstein and Erhlich 1989), this ancient brain is not attuned to recognizing climate change as a threat. As David Orr often says, climate change is not the Huns attacking us from over the ridge. Rather, its a slow process that more often than not does not pose any immediate danger to us. Moreover, people often think of it as a distant threat, both spatially and temporally, affecting people far away and perhaps affecting our com-

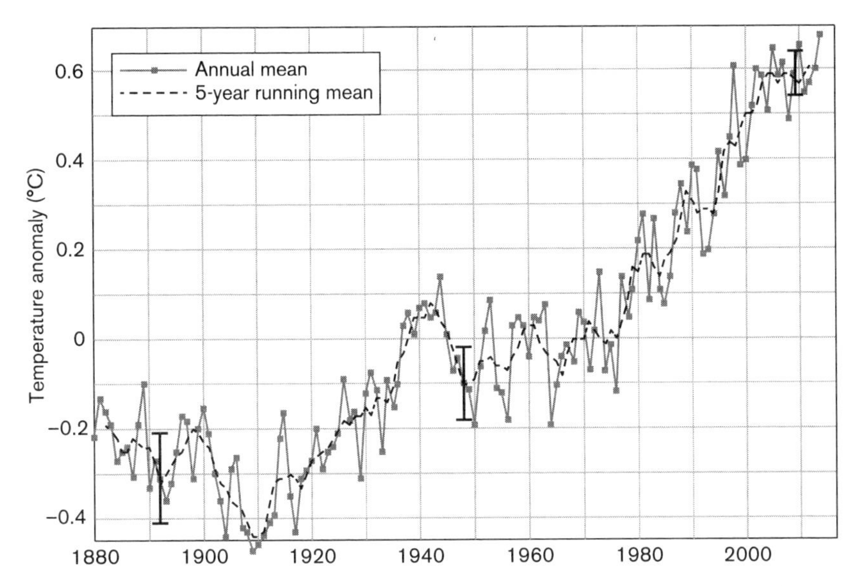

Fig 3.3. Temperature increases since 1880. (note for UCP: not on art log. Add with redraw submission)

munities in some far, distant time. Thus, it's not just a limitation of our senses, but a limitation of our ancient brain as well, that may lead us not to notice climate change.

This is not to say that everyone is oblivious to these changes. Individuals who spend a great deal of time outdoors notice them. For instance, while camping in Mesa Verde National Park in Colorado one spring, I had the chance to talk to a park ranger who had worked there for decades. He reflected on how the incidence of drought and major brushfires had increased during his time working at the park. Similarly, maple syrup producers in the northeast have noted how winters have shortened, leading them to harvest their maple syrup earlier and earlier (Kahn 2016). Bird-watchers have observed changes in migratory patterns as the world has warmed (Audubon Society 2014). These people's senses are tuned to nature. Their extensive time spent in nature has enabled them to take note of these changes.

URBAN LIFESTYLE / LIVING INDOORS While not everyone is oblivious to these changes, recall from chapter 2 that the individualistic worldview is strongly associated with industrialization, a life primarily lived in urban environments and spent indoors. That is, few of us frequent the outdoors like national park rangers, maple syrup producers, or bird-watchers. This lack of time spent outdoors may make us less aware of the changes occurring in the natural world. Thus, it's important to realize that something that accompanies an individualistic worldview is a life spent in certain places (more urban, more indoors) and not other places (outdoor, more natural settings). This lifestyle has implications for what we see, notice, and react to.

TECHNOLOGICAL DEVICES In this industrialized, technologically sophisticated culture we live in, the indoor/outdoor distinction has also lost some of its meaning. Namely, with the rise of smartphones and other portable electronic devices, people do not focus on computer screens, televisions, or other electronic devices indoors only; when outdoors, too, they may focus on portable devices. This brings up the question of the extent to which they notice their outdoor surroundings. As they walk along sidewalks, people increasingly seem to have their heads down while either texting or reading from their smartphones. Increasingly, as these devices become almost artificial appendages, what people view may be more and more limited to the screen protruding from their hand.

BEING SELF-ABSORBED The mastery-oriented individualistic worldview, with its emphasis on the "I," also inclines people to be self-absorbed. This self-absorption can prevent people from noticing events around them. The argument goes as follows. Attention is limited. Like the beam of a flashlight, it is directed only at certain things, and not all things at once. To the extent that it is focused on one's self, by definition it is not focused away from self, on others or nature. Consequently, when people do not notice events around them, their self-absorption can serve as a barrier to helping. I've conducted several studies examining the relationship between self-focus and helping, which illustrate this very point (Mayer, Duval, Holtz, and Bowman 1985; Rogers, Miller, Mayer, and Duval 1982). What was particularly interesting about this research, which dealt with prosocial behavior but fits so well with this discussion of proenvironmental behavior, is that high self-focus is most likely to undermine helping when the emergency is not salient. That is, when the emergency isn't that dramatic or noticeable (i.e., strong enough to draw an individual's attention away from herself or himself).

Given that the emergency inherent in CO_2 change and temperature change is of low salience (i.e., the conditions are not dramatic enough to capture the focus of attention), the self-absorbed "I" who characterizes the mastery-oriented individualistic worldview may be less inclined to notice these changes and, consequently, remain indifferent to the threats. So, the impact of the "I" may not be limited to a psychological disconnect of self from nature effect associated with stage 3 (i.e., a lack of "we"). Potentially affecting both stage 1 and stage 3, the emphasis on "I" associated with the mastery-oriented individualistic worldview further underscores the importance of this worldview as a barrier to caring for the natural world.

LACK OF PLACE ATTACHMENT Humans feel bonded to many different places (see Lewicka 2010 for a review). Among the places that people exhibit greater or lesser attachment to are their homes (Easthope 2004), neighborhoods (Kusenback 2008), cities (Tuan 1975), regions (Laczko 2005; Lewicka 2008), and countries (Reicher, Hopkins, and Harrison 2006). People also have varying amounts of attachment to natural places, such as lakes (Jorgensen and Stedman 2001, 2006), lake regions (Williams and Van Patten 1998), forests (Smaldone 2006), rivers (Davenport and Anderson 2005), wild streams (Hammitt, Backlund, and Bixler 2006), mountains (Kyle, Graefe, Manning, and Bacon 2003), seacoasts (Kelly and Hosking 2008), and landscapes (Fishwick and Vining 1992; Kaltenborn and Bjerke 2002).

Research shows that people who have a strong attachment to a natural setting exhibit greater awareness of the impact that humans have had on that setting. This awareness includes the effects of climate change. Like the bird-watchers and the maple syrup producers, people who experience place attachment notice changes occurring in the natural world. This research also illustrates how connection (a stage 3 consideration) affects stage 1 (i.e., noticing). Correspondingly, this greater awareness is also associated with a stronger desire to protect the setting (Stedman 2003).

Industrialization also influences attachment (i.e., the human bond with nature) through its impact on population mobility. Today, people rarely stay in the place where they were born and raised. For example, I've lived in ten different places in my life. Much of my movement has been associated with my education and work experience. Such movement also influences how we define ourselves: research indicates that increased mobility leads people to accentuate their personal attributes and traits rather than their social ties (Oishi, Lun, and Sherman 2007). This increased mobility has also been linked to weaker attachments to place (Lewicka 2011), with the corresponding lessening of awareness of changes to the natural world.

HABITUATION Another barrier to noticing climate change is habituation. After repeatedly hearing messages about climate change, people may eventually tune them out (Belch 1982; Burke and Edell 1986). It's much like a neighbor's too-loud stereo: you might initially orient yourself to the sound, but over time it may fade from your awareness—you won't even hear it.

ESCAPIST ACTIVITIES The human desire for escapist activities is another barrier that needs to be taken into account. During our lifetimes, we each face many emotionally weighty issues. Death, loss of employment, health issues, financial worries, and the trials of parenthood make up only a small fraction of stresses and strains that people face during their lives. For people working long hours, or for people simply trying to make it through each day as best they can, escapist activities can seem like a breath of fresh air. These activities provide a time to relax, let go of one's troubles, and zone out. Although following Cleveland sports teams is usually not very relaxing, I can escape to the next NBA title run for the Cavs, hope that this is the year for the Indians, or direct my attention to the next NFL draft for the poor Browns. To spend time thinking of climate change may be another burden in what already seems like a day, or at times a life, full of burdens. As a consequence, there are times when any of us may experience *a motivated not wanting to think about another ponderous issue*, which may include the issue of climate change.

GENERAL IGNORANCE In summary, ignorance of the problem is a barrier that keeps people from trying to solve it (cf. Gifford 2011). This ignorance is due to the limitations of our senses and our ancient brain. Our lifestyle, egocentric focus, and a motivated not wanting to think about this problem may also lead us to be oblivious to it. Additionally, people may hear about the threat so often that they finally tune these messages out.

Ignorance, however, can also be linked with the lack of a certain *kind* of knowledge. This is the less intuitive form of ignorance but no less important.

People guided by their values to climb the ladder to success and wealth, with their eyes fixed on the prize above them, may fail to notice that their aspirations negatively affect the natural world beneath them. These people may be highly knowledgeable. They may come from the best universities and colleges. The knowledge they have acquired, however, may be narrow and analytical. They may be bred within an academic discipline that does not nurture them to think of their actions within the context of the natural world. As is the case in many academic settings, the training is for employment and success in one's specific field of study. As Orr (1994) puts it, "The truth is that without significant precautions, education can equip people to merely be more effective vandals of the earth. If one listens carefully, it may even be possible to hear the Creation groan every year in May when another batch of smart, degree-holding, but ecologically illiterate, *Homo sapiens* who are eager to succeed are launched into the biosphere" (p. 17). More education can, ironically, be associated with noticing less and harming more. Thus, ignorance is not always about the absence of information, for the presence of information can also blind people to the world around them. This is an issue that we'll return to when considering stages 4 and 5, for ignorance also influences people considering what actions to take and how to actually perform specific actions.

Stage 2. Interpreting the Event as an Emergency

First, an event must be noticed, but noticing something is a far cry from reaching out and helping. Stage 2 emphasizes that the path to helping requires people not merely to see the event but to interpret the event as an emergency. What factors might lead them to see it as an emergency or not? As was the case with stage 1, multiple factors influence people at this interpretive stage. One critical point to remember as you read this section is that many emergencies are *ambiguous*, and there is *uncertainty* associated with the event. More often than not, the emergency does not entail blood spurting from someone's artery, bombs exploding, or some other clear-cut, unequivocal event.

INCONGRUITY OF PLEASANT WEATHER AS THREAT For instance, consider the paradox of sounding an alarm over a pleasant event. Imagine wearing sunglasses and T-shirts in January in a place where historically temperatures

BOX 3.3. STAGE 2. BARRIERS TO INTERPRETING THE EVENT AS AN EMERGENCY

- Incongruity of pleasant weather as a threat
- Optimism bias
- Technosalvation / suprahuman powers
- Discounting the messenger / mistrust
- Right-wing political ideology
- Disinformation campaigns
- A general atmosphere of uncertainty
- Social comparison
- Selective exposure
- Social norms / pluralistic ignorance
- Denial
- Psychological distance (hypothetical, space, time, or social distance)

averaged in the teens. This event is becoming more frequent in the northern latitudes. How are people expected to see an emergency when they're feeling happy? Emergencies are typically associated with distress, discomfort, cries in the night, sirens blaring. Living in a northern climate, where winters can often be brutally cold, I feel awkward when I tell people that they should be worried about an atypical warm and pleasant winter or an unexpectedly warm and pleasant winter day. Getting people alarmed about a seventy-degree day in the heart of winter is difficult. Where I'm from, people are rejoicing at our good luck rather than bemoaning the fact that it's not ten degrees outside. In their view this is not a symptom of an emergency.

OPTIMISM BIAS Additionally, Americans tend to be an unusually optimistic and confident group of people. This bias has been defined as the psychological tendency of people to think "they are less likely to experience negative future events and more likely to experience positive future events as compared to others" (Wei, Lo, and Lu 2007, p. 667). Optimism has often been considered one of our great strengths as a people (Keller 2015). However, a person's greatest strength can sometimes be the person's greatest weakness. When optimism is overdone, it can lead people to discount their personal risks for a heart attack (Weinstein 1980). It can also lead people to discount environmental risks, such as radon exposure (Weinstein, Klotz, and Sandman 1988) or other environmental hazards (Hatfield and Job 2001; Pahl, Harris, Todd, and Rutter 2005).

In the case of climate change, feeling optimistic and confident may lead people to not worry about it. Optimism and confidence can lead people to think that while the risks from climate change will in all likelihood worsen over the next twenty-five years, the risks are worse for people living in other places and not where they live (Gifford, Scannell, Kormos, Smolova, Biel, Boncu, Corral, Guntherf, Hanyu, Hine et al. 2009).

Moreover, the optimism and confidence that derives from the United States' worldview can lead people to minimize the severity of what it will take to deal with the threats. Military history provides several telling examples. For instance, at the advent of World War I, both Germany and Great Britain, on the basis of their feelings of optimism and confidence, estimated that the war would end quickly. As Stoessinger (2010) states in his book *Why Nations Go to War*: "The emperors and generals who sent their men to war in August 1914 thought in terms of weeks, not months, let alone years." Similarly, at the beginning of the Civil War, on the basis of an overinflated sense of optimism, both sides anticipated a brief conflict (McPherson 2003). Related to the threat of climate change, people may well see this threat as simply another obstacle that we'll readily overcome. Once we put our minds to it, we'll meet the challenge with American know-how, just as we did when putting men on the moon. From this perspective, the emergency isn't nearly as ominous as many would have us believe. Optimism and confidence can lead people to deny that an environmental emergency even exists.

TECHNOSALVATION / SUPRAHUMAN POWERS Optimism can lead people to believe that someone will certainly invent a "silver bullet," a technological innovation that will take care of the problem. For others, their observation that technology is advancing at a rapid pace may persuade them that it is only a matter of time before a technological innovation occurs that will make everything right again. After all, researchers are making progress in developing various forms of carbon capture in order to cleanse the skies of excess CO_2. Two technologies being developed entail using artificial trees and biomass to extract CO_2 from the atmosphere. The fact that advances in medical technology have saved countless lives may persuade people to place their faith in technosalvation and, as a result, feel less distressed about climate change.

Other people place their faith in a religious deity, or in a secular deity such as Mother Nature. These individuals have faith that this suprahuman power will save us. Or they may think this higher power will act in a manner beyond our control, so worrying has very little purpose. In an example provided by Gifford (2011), "researchers who interviewed two groups of Pacific Islanders who live on very low-lying atolls threatened by rising sea levels found that one group is already purchasing higher ground in Australia; the other group, trusting that God will not break the Biblical promise never to flood the Earth again after the flood that Noah and his entourage endured, believes that sea level rises will not affect them because there will be 'fire next time' (Mortreux & Barnet, 2009)."

DISCOUNTING THE MESSENGER / MISTRUST This embrace of optimism and confidence can lead people to reject not just the message that environmental threats exist but the messenger as well. Climate scientists and environmentalists who warn people of environmental threats can be viewed as pessimists. In some people's minds, what is being said becomes more a reflection of the person saying it than of something occurring in the world. Let's think about this for a minute. When medical doctors tell patients that they have cancer, the patients never think of telling the doctors that they're simply being pessimistic. Patients never think that what they are being told is a reflection of the medical doctor rather than something going on in their own bodies. Yet when climate scientists and environmentalist warn people of environmental threats that are every bit as real as the cancer in a patient's body, it is not that unusual for people to see the problem as located in the scientists (i.e., they're "doom and gloomers," or they can't be trusted) and not in the world.

Anti-intellectualism is another factor that leads to discounting the messenger. The characteristic of anti-intellectualism, which can be traced back to ancient Greece, has been present in the United States worldview from colonial times (Sowell 2001). More recently, the use of the terms *eggheads*, *geeks*, and *nerds* illustrates Americans' general distrust of intellectuals. Intellectuals themselves tend to be viewed as different and can be socially rejected. Consequently, when intellectuals claim a threat is present, their claims may often be dismissed.

For example, during the 1952 presidential race, Republican vice-presidential candidate Richard Nixon used the slang term *egghead* to describe Democratic presidential candidate Adlai Stevenson. It was a derogatory term meant to dismiss his arguments, a way to characterize him as out of touch with the ordinary person, as lacking down-to-earth realism and common sense. In a Pulitzer Prize-winning book, *Anti-intellectualism in American Life*, the historian Richard Hofstadter (1963) documented the basic mistrust of intellectuals by many people in the United States. More recently, Susan Jacoby's book *The Age of American Unreason* (2009) updates Hofstadter's work and illustrates how anti-intellectualism remains an important current in American thought.

How does this mistrust of intellectuals relate to stage 2? Given the limitations of our senses to detect CO_2 and temperature changes, the message that these changes pose a threat is largely being conveyed by scientists (i.e., intellectuals). Atmospheric scientists and scientists who specialize in climate modeling, who forecast future changes, are sounding the warnings. If a person doubts the legitimacy of these scientific appraisals, he or she is likely to not see an environmental emergency associated with climate change.

RIGHT-WING POLITICAL IDEOLOGY In the United States, right-wing political denial of climate change and opposition to mitigation policies has also been linked to a mistrust of environmentalists. With the fall of the Berlin Wall and the communist threat, some have argued that right-wing adherents have positioned environmentalists as the new opposition group to fear and deride (Jacques, Dunlap, and Freeman 2008). The "Green Scare" has now replaced the "Red Scare," and the right-wing distrust of environmentalists is nicely captured by their characterization of environmentalists as "watermelons," green on the outside but red (communist or socialist) on the inside (Hoffarth and Hodson 2016).

As the right-wing commentator Charles Krauthammer (2009) has stated, "With socialism dead, the gigantic heist is now proposed as a sacred service of the newest religion: environmentalism. . . . The Left was adrift until it struck upon a brilliant gambit: metamorphosis from red to green." This characterization of environmentalists leads right-wing adherents to view them as having ulterior motives aimed at threatening the American way of life. Some view the environmentalists' use of climate threat simply as a ruse to achieve their true aim (greater government control) and as a threat to free-market capitalism. Thus, this distrust of environmentalists' motives leads them to disregard the message that a climate emergency exists.

It is not just that right-wing adherents disregard the environmentalist message, however. They also stridently oppose what they see as environmentalists' attempts to subvert the status quo. As reported by Mooney (2015), Rush Limbaugh, a popular right-wing radio celebrity, in reaction to the environmentalist request to turn off lights during Earth Hour, told his audience, "I wanted to make sure I could use as much damn power as I could. . . . I turned the thermostats down to 70 degrees, 68 degrees [during hot weather]. I turned on every light in the house! I turned on every light in the back yard and aimed 'em down so they wouldn't hit the turtles! I mean, I had my house lit up like a Christmas tree last night." This opposition isn't simply from Limbaugh. When considering investing in energy-efficient technology and purchasing "environmentally friendly," energy-efficient lightbulbs, right-wing adherents are more resistant (Gromet, Kunreuther, and Larrick 2013). Thus, it isn't simply the case that right-wing adherents reject the environmentalist message. There is a boomerang effect, where they go out of their way to do the exact opposite of what is called for. As Mooney (2015) states, Limbaugh "thumb[ed] his nose at environmentalists."

Another reason why right-wing adherents are less concerned about climate change than liberals has to do with their concern with maintaining the status quo. That is, they may be more concerned with maintaining a social order that benefits their group in some way (Jost and Benaji 1994; Jost, Nosek, and Gosling 2008). For instance, right-wing adherents are more likely to embrace the self-enhancing belief in human supremacy over nature. Moreover, in their pursuit of success, they are more likely to believe that nature exists to be exploited by them (Hoffarth and Hodson 2016). Maintaining the status quo also involves protecting free-market ideology. When solutions to climate change focus on constraints on free-market capitalism, Republicans were found to be particularly likely to deny climate change (Campbell and Kay 2014, study 2). Right-wing ideology places the value of the economy and jobs over considerations for protecting the environment (Hoffarth and Hodson 2016).

Right-wing adherents may also differ from liberals in terms of the relative importance they place on five core moral domains (Graham, Haidt, and Nosek 2009; Haidt and Graham 2007). The issue here concerns what is worthy of moral concern. What has been shown to differentiate conservatives from liberals is the former group's greater emphasis on the moral domains of in-group loyalty (favoring one's in-group over an out-group, which has been related to patriotism), authority (a desire to maintain the status quo and to obey authority), and purity (being opposed to a hedonistic lifestyle and placing a value on chastity). A sixth moral domain, economic liberty (protecting the free market, privileging the economic rights of the individual to, through hard work, succeed, along with opposition to government intervention and wealth redistribution), has also been proposed as a main concern for conservatives (Rossen, Dunlop, and Lawrence 2015).

Liberals, on the other hand, place a greater emphasis on two moral domains: harm (a threat to the well-being of individuals) and fairness (a violation of an individual's rights). That climate change may be more strongly linked to the moral domains of harm and fairness (i.e., climate change in all likelihood negatively affects the most vulnerable in the world first), than to the moral domains of in-group loyalty, authority, and purity, has led some researchers to argue that this may at least partly explain why right-wing adherents are less concerned about climate change. And if the climate-change message emphasizes harm done to the most vulnerable, this type of message might motivate liberals, but not conservatives, to take action. Moreover, if the message contains plans for government intervention, conservatives might view the proposal as a threat to the status quo or to the free market. This would lead to their moral opposition.

DISINFORMATION CAMPAIGNS The strident opposition of right-wing adherents and others interested in protecting the status quo also comes in the form of disinformation campaigns. There are reports of how oil tycoons, fossil fuel companies, and others with a vested interest in maintaining their lucrative business interests have channeled millions of dollars to groups whose aim is to cast doubt on climate change. It has been reported that between 2002 and 2010, nearly $120 million was provided to different groups to basically befuddle the issue by trying to discredit climate science (Goldenberg 2013). Given the widespread tendency to distrust intellectuals, the messages from right-wing conservatives, and the disinformation campaigns by powerful interest groups, is it really any wonder that many people have doubts about whether climate change is an emergency?

A GENERAL ATMOSPHERE OF UNCERTAINTY As to whether climate change is an emergency, a general atmosphere of uncertainty arises in several ways. First, as mentioned earlier, climate change is not a 9/11 event. Second, given that people are generally optimistic, many may simply think that the environmentalists are overreacting to a problem already being solved. Third, given a general mistrust of intellectuals, right-wing conservative mistrust of "watermelon" environmentalists, and disinformation campaigns, in the public news arena, there seems to be no clear consensus about whether climate change is an emergency. Fourth, scientists, as a matter of course, always state their claims with a margin of uncertainty. This may inadvertently lead many people to view climate change as less of an emergency than scientists intended (Budescu, Broomell, and Por 2009).

SOCIAL COMPARISON Given uncertainty, Festinger's (1954) social comparison theory argues, people will attempt to reduce this uncertainty. How might they do this? Who are they likely to seek out? In all likelihood they'll seek out either people they see as similar to themselves or authority figures they trust. In many instances, then, uncertainty may lead people to seek clarifying information from people who hold a mastery-oriented individualistic worldview. In other words, they may seek out information from people who have not noticed this threat or who stridently oppose the climate-change message. Some may turn to a conservative talk-show host like Rush Limbaugh, conservative politicians, or esteemed conservative religious leaders. For a wide variety of reasons, then, people may not see climate change as an emergency that needs to be addressed.

SELECTIVE EXPOSURE Once people have reduced their uncertainty, their beliefs become resistant to change. This is especially true if a person has become committed to the belief by having freely chosen to say something about, or act in manner supportive of, the belief (Mayer, Duval, and Duval 1980). According to Festinger's (1957) cognitive dissonance theory, people under conditions of commitment are not rational, but rationalizing, animals. Once committed to a position, they are motivated to maintain consistency between their beliefs and behaviors. When acting in a manner not reflective of these beliefs and behaviors, they are thought to experience cognitive dissonance, an unpleasant cognitive state. This negative state motivates people to reestablish and maintain consistency.

How might they maintain consistency? Several strategies have been documented. First, people can seek out information that confirms their views, an act called selective exposure: when people have one set of beliefs and are provided with two packets of information, one that supports their beliefs and one contradictory, they are likely to accept the packet that supports their beliefs. Or, when selecting television stations to watch, liberals may well be more likely to selectively expose themselves to the liberal MSNBC than to the conservative FOX news program, while conservatives may do exactly the opposite.

From a cognitive dissonance perspective, if people do not believe that an environmental emergency exists, they may not read this book. They may not seek out the 350.org website that provides information that would contradict their beliefs. They may skip over articles in the newspaper in which scientists present new information about how climate change is occurring more quickly than predicted. On a night out, they may not go to the movie *Chasing Ice*, which documents glaciers receding at an alarming rate, and so on. Besides avoiding exposure to such information, they may also rationalize the behavior of the "so-called experts" as simply that of "eggheads" or liberal-minded individuals who have an ulterior political agenda. In other words, there are many ways that people can filter and construe things in order to maintain and bolster their prior beliefs.

Regarding stage 2, then, there are many barriers that prevent people from seeing climate change as an emergency. The image of human nature I'm painting here *is not* that of the rational person open to information, weighing the strengths of various arguments, and coming to a logical decision. Rather, more in line with Festinger's cognitive dissonance theory and social comparison theory, what emerges is an image of people looking to trusted others to help them understand their world—people seeking stability and consistency in their views by associating with, listening to, and reading articles written by like-minded people. Among such individuals, appeals to reason alone may fail; in particular, appeals by someone who is viewed as dissimilar may fall on deaf ears.

SOCIAL NORMS / PLURALISTIC IGNORANCE In one's social life, the values that characterize a nation provide guidelines for appropriate behavior. For example, hard work is something we value, and this value becomes manifest in normative expectations that people will work hard and not be lazy. Just as a hallway and doors in a home constrain where a person can walk, social norms also constrain behavior. When acting in a socially accepted manner, people will not face social resistance. In contrast, a person who acts contrary to norms will walk into a wall of social criticism. Generally speaking, people want to appear competent and capable. They desire to walk in the hallways and enter the rooms where social praise awaits. People are often wary of saying things they anticipate will receive negative social sanction. We avoid certain conversations. We often remain silent when we anticipate that others might disapprove of what we have to say.

It's been observed that only one in four Americans report that they regularly discuss the issue of climate change (Leiserowitz, Maibach, Roser-Renouf, Feinberg, and Rosenthal 2015). Geiger and Swim (2016) argue that this lack of discussion is a barrier to change, for "interpersonal communication about topics is crucial to build public acceptance and support for social change: scientifically grounded public discussion can increase public understanding of the problem, community engagement, and development of consensus for locally appropriate mitigation and adaptation solutions (Clayton et al., 2015; Swim, Fraser, and Geiger, 2014)." Why this lack of discussion? Geiger and Swim suggest that there is a *socially constructed silence* on this issue that results from pluralistic ignorance. Pluralistic ignorance refers to the tendency for a majority of people to misperceive other's beliefs on an issue. They inaccurately think that fewer people share their viewpoint than actually do (Prentice and Miller 1993). In two studies, Geiger and Swim found that people feel hesitant to discuss climate change because they wish to avoid the criticism of others. They don't want to feel less competent. They didn't want to risk losing the respect of others. Their findings highlight the fact that people often falsely perceive the social norm as one that rejects, as inappropriate, talking about climate change. This silence results in an ominous hush, for the lack of discussion prevents people from reaching a consensus that climate change is an emergency that demands action.

DENIAL We've already discussed a variety of factors that lead people to deny the threat of climate change. From a motivated not wanting to think about an issue to the discounting of a message owing to mistrust of the messenger, denial can take many forms. However, denial can also result from fear. Feeling incapable of doing anything in the face of the catastrophe that looms before us is a nightmare scenario. In the same way that people's fear of their own mortality can lead them to deny the inevitability of death (Solomon, Greenberg, Schimel, Arndt, and Pyszczynski 2004), climate change presented as a nightmare scenario can increase people's fear of their own mortality, prompting them to deny that it is an emergency demanding our attention.

Years ago I began documenting how a permanent intraself discrepancy (i.e., a permanent difference between who a person is and who she or he aspires to be) can lead people not just to denial but also, in some contexts, to an active and angry response to those who bring up an unwanted topic (reported in Duval and Duval 1983; see Duval, Mayer, Duval, and DePould 1980). The argument my colleagues and I made is that when someone has a permanent intraself discrepancy, that person will experience pain. Not wanting to remain in this permanent painful state, the person will be motivated to avoid thinking about it (i.e., to deny it). However, if some other person reminds the individual of this pain (i.e., "brings it up"), this other may be seen as the cause of the pain because he is contiguous with the experience of the pain. As a result, the denier will direct pain at this other person in the form of an anger response. In the same way, a person with a permanent intraself discrepancy who is confronted by both the threat of climate change and an inability to reduce the threat may not only deny that an emergency exists but also vehemently derogate a messenger who reminds her of this painful topic. Denial, then, is not simply a passive response from people who don't talk about, or don't want to think about, an issue. As we've seen before, it can also lead to active, angry denunciations.

PSYCHOLOGICAL DISTANCE (HYPOTHETICAL, SPACE, TIME, OR SOCIAL DISTANCE) Another factor that contributes to stage 2 is psychological distance, or how far climate change is removed from the self. As a concept related to, yet distinct from, place attachment, psychological distance is an obstacle to helping when people are uncertain about whether climate change exists (hypothetical distance) or think it will occur only in a distant place (spatial distance), in the distant future (temporal distance), and to people who are dissimilar or not part of their own community (social distance).

An interesting review of the literature by Rachel McDonald, Hui Yi Chai, and Ben Newell (2015), however, paints a more nuanced picture. These authors make the point that, while decreasing psychological distance has been shown to increase *concern* about climate change, portraying it as an event that is too near may prompt some people to help but others to adopt denial. Political ideology comes into play, too. In the case of conservatives—who, as discussed earlier, place greater value on their in-group than on out-groups—*decreasing* social distance (i.e., providing them with information demonstrating that climate change will harm their group) tends to make them more concerned and willing to do something about it. In contrast, liberals—who are more concerned about harm and fairness—strongly respond to *increasing* social distance (i.e., portrayals of climate change as harming people in developing nations) and become more likely to help. Thus, an interesting implication of this research concerns environmental organizations and their information campaigns. Messages should be tailored to the groups they are directed to.

Stage 3. Feeling Personal Responsibility and a Sense of "We"

Recall that a sense of personal responsibility can arise primarily in two ways. First, it can arise from seeing yourself as having caused the problem. For instance, if I accidently spilled a glass of water on someone, I'd apologize for having caused the problem and I'd help the person in some way. In this

BOX 3.4. STAGE 3. FEELING PERSONAL RESPONSIBILITY AND A SENSE OF "WE"

- Factors influencing a sense of "we"
  - Mastery-oriented individualistic worldview
  - Lack of place attachment
- Factors influencing causal attributions
  - Diffusion of responsibility
  - Temporal/spatial proximity
  - Self-enhancing tendencies
  - Just-world beliefs
  - Ethnocentric beliefs
  - Magnitude of cause and effect

section, I highlight factors that affect perceived causality as related to climate change. Second, as stated previously, consciousness of personal responsibility can arise from feeling a sense of "we" in relation to either another person or nature. This sense of connection promotes both personal responsibility and caring for others and for the natural world.

MASTERY-ORIENTED INDIVIDUALISTIC WORLDVIEW From my perspective, the impact of this worldview on proenvironmental behavior cannot be emphasized enough. As mentioned previously, this worldview, with its emphasis on "I" rather than "we," undermines caring for the natural world. Something I haven't covered, however, is that in the United States this "I"/"me" orientation seems to be increasing. An interesting study by Jean Twenge, Keith Campbell, and Brittany Gentile (2012) examines this trend by looking at pronoun use in a Google database that included 766,513 American books published between 1960 and 2008. They found that during this time, the use of first-person singular pronouns ("I"/"me") increased by 42 percent, while the use of first-person plural pronouns ("we"/"us") decreased by 10 percent and second-person pronouns ("you"/"your") showed a 300 percent increase. This trend is worrisome.

It's worrisome for several reasons. First, it's worrisome because an increasing emphasis on "I" instead of "we" may permit people to feel even less obliged to engage in proenvironmental behavior. Second, it suggests that the human/nature split that is thought to be the fundamental reason why people remain indifferent to the harm done to nature may only be increasing. Ultimately, however, this trend is worrisome because the time we have before reaching a tipping point is running out, and the greater emphasis on "I" indicates that we're not moving in the right direction to avoid this tragedy.

LACK OF PLACE ATTACHMENT As noted earlier, the lack of place attachment affects stage 1. Now let's consider how it also relates to stage 3. The population mobility associated with an industrialized world, and people's greater tendency to move from place to place, weakens a person's attachment to any one place. Moreover, the long hours at work, the overall amount of time spent indoors, and the increasing length of time spent focused on electronic devices only adds to a person's sense of disconnection from nature. Feeling connected to someone or to nature involves time—time spent with others or in the natural world. A sense of "we" doesn't magically appear. Much of modern life discourages building these stronger ties.

FACTORS INFLUENCING CAUSAL ATTRIBUTIONS Besides a sense of "we," there are other factors that influence individuals' sense of personal responsibility—namely, factors related to causal attributions encourage personal responsibility and helping behavior. When discussing the causal attribution literature, two points are important to keep in mind. First, realize that psychologists primarily study factors that lead to perceived causality (i.e., factors that give rise to individuals' subjective experience of what is causing what) and not objective causality (i.e., not what is actually causing an event to occur; see the perception-versus-reality discussion in chapter 1). Second, when studying the factors that influence perceived causality, recognize that a variety of factors have been shown to influence whether we see ourselves as causally related to an event. What's important to consider here is the *interplay* of these different factors. Under some conditions one factor may override another factor, while under other conditions some other factor may contribute more to a causal impression.

DIFFUSION OF RESPONSIBILITY Recall that a major factor emphasized in Latane and Darley's research is diffusion of responsibility. With millions of people present who are potentially able to help avert further climate change, any one person may feel only minimally responsible for providing aid. Moreover, with millions of people contributing to climate change, diffusion of responsibility may lead people to view themselves as only minimally responsible for having caused the problem in the first place. Both of these considerations may lessen the likelihood of a person helping.

TEMPORAL/SPATIAL PROXIMITY Another major factor influencing perceived causality is the temporal and spatial proximity between a behavior and an outcome. When I flip a light switch (the behavior), I see a light go on (the event). If the light goes on immediately after the behavior, this is said to be an instance of high temporal proximity, and I will form a strong causal impression. If I flip the switch and the light goes on a minute later, this lower temporal proximity will weaken or nullify the causal impression. As for spatial proximity, if I flip a switch and a light goes on near me, I'll form a stronger causal impression than if I flipped the switch and saw a light go on some distance away. Given these two factors and their interplay, we can say that the strongest causal impression will occur when I flip the switch and immediately a nearby light goes on, while the weakest impression will occur when I flip the switch and there is a delay before a more distant light goes on.

SELF-ENHANCING TENDENCIES When temporal and spatial proximity is high, other factors influencing causal impressions may play smaller roles. For instance, when you're "caught with your hand in the cookie jar," the general self-enhancing tendency to associate positive outcomes with self and negative outcomes not with self may not come into play. In this example, the strong temporal and spatial proximity would make it difficult for someone to say, "It wasn't me." However, when proximity is absent or low, self-enhancing or even group-enhancing attributional biases may be more likely to influence the result. Under these conditions, one may be more likely to dissociate a negative event from oneself or one's group, which helps maintain the general favorable impression one has of oneself or one's group.

As is so clear, the temporal proximity of both our CO_2-emitting actions and the effect of climate change is anything but proximate. The temporal proximity is low: the time lag between these actions and the impact they have on the climate is years. Similarly, the spatial proximity between our CO_2-

emitting actions and atmospheric effects is low. CO_2-emitting actions of people in the Midwest might have contributed to a megastorm on the East Coast, or anywhere in the world, for that matter. These actions will not lead to the spatially proximate occurrence of an atmospheric disturbance, such as a tornado, in one's backyard. Together, the lack of temporal and spatial proximity between CO_2-emitting actions and the effects of climate change make it difficult for people to form the impression that they are causally responsible for these climate effects. This is not a case of flipping a switch and a light going on.

JUST-WORLD BELIEFS The fact that the contiguity between these actions and climate change is low may also open the door for other factors, such as just-world beliefs, to influence individuals' causal attributions. If we consider ourselves to be "good" people, then it follows that "good" people do "good" things and not "bad" things. In relation to the negative environmental threats before us, this factor may make it difficult for people to see how they or the valued groups they belong to could be responsible for negatively affecting the natural world.

ETHNOCENTRIC BELIEFS Given the lack of temporal and spatial proximity, the ethnocentric beliefs of a nation—which are another related factor, aside from just-world beliefs and self-enhancing factors—may also come into play. The ethnocentrism associated with a nation (i.e., people viewing their nation as special) does not fit with seeing one's group in an unfavorable light. In a sense, this fitting of a positive (our nation is special) with a negative (our nation is one of the prime contributors to climate change) is like fitting a square peg in a round hole. This apparent lack of fit may prevent people from taking responsibility for the negative event of climate change.

MAGNITUDE OF CAUSE AND EFFECT The magnitude of the causal event and the effect event is another factor to consider. Stated differently, small-magnitude causes are unlikely to be viewed as causally related to large-magnitude effects. For example, it's unlikely that a sneeze will be viewed as causing a garage to blow down. The magnitudes of the two events simply do not match. Similarly, it might be just as difficult for people to view the small event of starting their car engine, and the resultant CO_2 emissions, as in any way related to the megastorms associated with climate change. But the

BOX 3.5. STAGES 4 AND 5. BARRIERS TO FORMING AN IDEA OF WHAT TO DO AND HAVING THE ABILITY TO DO IT

- Feeling overwhelmed
- Ignorance
- Fatalism
- Perceived and collective efficacy
- Depleted directed attention
- Lack of creativity
- An emphasis on cleverness and analytic thinking
- Habit/commitment
- Rebound effect
- Mistrust

accumulation of many drops of rain can cause a flood, and the accumulation of CO_2 emissions from many people's cars can add up to a major effect. This view that cumulative actions cause large effects needs to be impressed upon people. However, in an individualistic culture, where individual, and not collective, action is emphasized, trying to impress upon individuals the influence of collective action may be difficult, especially when self-enhancing factors are in play.

Stages 4 and 5. Forming an Idea of What to Do and Having the Ability to Do It

FEELING OVERWHELMED Once we have progressed through stage 3 of the model, the issue becomes: what is a person to do, and is the person able to do it? Given the global nature of the climate-change threat, it is easy to feel overwhelmed. The Greenland ice sheets are melting; the oceans are warming, resulting in stronger hurricanes; and the western United States is experiencing another drought with accompanying massive wildfires. What can a person do? Is it any wonder that at times an individual may be tempted to simply turn on the television and stop thinking about it?

I can remember being a kid at the L.A. airport waiting with my family for a flight. My father insisted on arriving at the airport three hours ahead of time, so we always had a great deal of time on our hands before our departure. In this instance I was watching people pass by, when a middle-aged man had a seizure and collapsed to the floor. I simply stared. I had no idea what to do. I remember the sound of his head hitting the floor like an egg hitting the side of a frying pan. I didn't feel well. I certainly noticed the event. Clearly, it was an emergency. As for feeling a sense of personal responsibility, I was near enough to feel as if I ought to do something, but I also felt overwhelmed. It was all too much for me. I wonder whether at times the environmental threats we face are similarly too much for any one person to deal with.

IGNORANCE In the airport example, the problem was more than simply my feeling overwhelmed. I didn't know *what* to do. I knew something could be done, but personally I was at a complete loss. Ideas cascaded through my brain, but none of them seemed appropriate. Even when people who think about climate change believe "the boulder" is already tilting and ready to roll down the hill, they may not have sufficient knowledge to form a plan. Alternatively, they may think there is an answer, such as developing forests of artificial trees to absorb CO_2, but may not know how to put the idea into action. Thus, even if you form an idea, if you can't enact it, you will not help.

FATALISM At other times, however, people may simply throw up their hands and say, "What will happen, will happen." They may conclude that there is simply nothing they or anyone else individually or collectively can do to confront the threat of climate change (Lorenzoni, Nicholson-Cole, and Whitmarsh 2007; O'Connor, Bord, and Fisher 1998). This is the person sitting in the dining room, puffing on a Cuban cigar, and sipping on a final glass of Scotch as the *Titanic* sinks.

PERCEIVED SELF-EFFICACY AND COLLECTIVE EFFICACY A major factor of stages 4 and 5, then, is individuals' perception that either individually or collectively they can come up with an effective plan to counter climate change and enact it. Self-efficacy is a critical issue. When people feel they can do something about the threat of climate change, they are less likely to engage in defensive denial and the derogation of opposing viewpoints (Fritsche, Cohrs, Kessler, and Bauer 2012). Instead, they are more likely to actively engage in proenvironmental acts. Given the scale of the climate change threat, a growing body of research focuses on collective efficacy. It may well be the case that only through collective efforts do we have any chance of effectively dealing with this threat.

A major obstacle to collective efficacy is that our sense of community has weakened over the past half century (Putnam 2001). Increasingly, as the pronoun-use study by Twenge, Campbell, and Gentile (2012) illustrates, we are more of an "I"/"me" culture than a "we"/"us" culture. Hyperindividuality may undermine our sense of being part of a collective, and this in turn may undermine any sense of collective self-efficacy that we can attain. Nevertheless, external threats can inspire people to strongly identify with a group, and if that group is actively engaged in trying to mitigate the impact of climate change, this can increase collective efficacy (Stollberg, Fritsche, and Baecker 2015). Even if the climate threat is not made salient, heightening people's awareness of a group to which they belong, and which is engaging in effective climate action, can increase people's sense of collective and personal self-efficacy (Jugert, Greenway, Barth, Buchner, Eisentraut, and Fritsche 2016). Building strong communities that effectively confront the threat of climate change is a critical task that belongs at the forefront of every climate activist's mind.

DEPLETED DIRECTED ATTENTION Steve and Rachel Kaplan argue that many aspects of our modern, urban life turn people into less effective problem solvers. They argue that people need directed attention in order to solve problems, but that this form of attention is limited, often depleted by the constant effort entailed in navigating urban centers, dealing with work, and other personal tasks that occupy a person's daily life. Moreover, besides being less effective at problem solving when their directed attention is depleted, people also become more irritable, aggressive, and less cooperative. This line of reasoning and the research that supports it (Kaplan and Kaplan 1989, 2003; Kuo and Sullivan 2001; Berman, Jonides, and Kaplan 2008) suggest that modern life may make us less effective in resolving the climate-change threat. This work also suggests that we may be less effective in confronting climate change because modern life has led us to be less cooperative with one another, which would undermine collective efforts to confront the threat.

LACK OF CREATIVITY Another aspect of modern life that influences our problem-solving ability is our emphasis on material rewards. Teresa Amabile's (1996, 1983) experimental research has shown that extrinsic motivation undermines creativity, while intrinsic motivation nourishes it. Given that creative solutions will be critical when considering how to mitigate the climate-change threat, our culture's emphasis on material rewards may undermine our ability to effectively plan actions that can lessen that threat.

AN EMPHASIS ON CLEVERNESS AND ANALYTIC THINKING The distinction between cleverness and wisdom is important to consider because we tend to value cleverness, often with little consideration for wisdom (see Sternberg 2012 for a similar discussion). For example, my dad told me that as a young man beginning his career as an aeronautical engineer in the 1940s, he once consulted on the effort to reduce the weight of the H-bomb. As initially developed, it weighed too much for a plane to carry. In a clever approach, he changed the bomb's casing, specifying a lighter substance. A clever solution, but was it wise? Similarly, developing the H-bomb was clever, but the cloud it has cast over humanity since its invention has led many to question the wisdom behind its development.

In another instance, I attended a psychology convention and came upon a talk concerning the intelligence of nations. Looking at one of the visuals, I noted that the United States and western Europe scored high on intelligence (cleverness). In my mind, I also mapped onto the image the historic and present CO_2 contributions from those nations. I thought about how the cleverness of these nations has led us into the progress trap of climate change. Generally, when considering how to get out of this trap, I wonder whether a nation that seems to emphasize cleverness over wisdom can effectively escape.

It is also important to evaluate proposed ideas to escape this trap in terms of whether the solutions are analytic or holistic. It is unlikely that a problem that derives at least partly from analytic thought will be resolved by using a similar style of thinking. Given that the challenges we face are due to relational issues (i.e., our general embeddedness in the natural world), more holistic solutions are called for. Piecemeal solutions in all likelihood will fail.

HABIT/COMMITMENT Another obstacle to change is the fact that our habits and commitment to a previous lifestyle may make change difficult. It is often said that the best predictor of future behavior is past behavior. Our habit of doing things a certain way, just like our general behavioral orientation (i.e., the general road we've been running on), leads in a direction opposite to the newly proposed direction. It may not be easy for people to change course. As they say, old habits die hard.

REBOUND EFFECT Even when we swear off the old ways and start in a new direction, there is the question of how long it will last. Among habits ranging from dieting to quitting smoking, an initial change of habit doesn't guarantee a permanent change of behavior. People often return to their old ways. Similarly, when we engage in more eco-friendly behavior, the change may last for a while, but the comfort of running on the old road may be too great to resist.

MISTRUST Lastly, when we consider developing ideas and a plan of action, and we can't come up with the solution, someone else may place one before us. Who is this person, though? Is the person a trusted source? Is the person someone you can identify with and believe in? I talked about mistrust earlier, and the same issues pertain to stages 4 and 5: if you mistrust someone, why would you have faith in the person's idea and follow this person in a new course of action?

THE OVERRIDING COST/BENEFIT ANALYSIS

This analysis is an overarching concern that potentially can come into play at any time within any of the five stages of Latane and Darley's model. The following potential costs might persuade someone to not get involved or to cease involvement.

RISKS For instance, a person might feel a greater *physical risk* by driving a small eco-friendly car rather than a less eco-friendly SUV. My daughter drives a small car to work. In the winters, with the winds blowing and snow on the streets, she doesn't feel safe and at times is actually scared. She is presently seriously considering buying a heavier, more substantial, four-wheel drive compact SUV. Similarly, while riding a bicycle is an eco-friendly mode of transportation, when I'm in a city watching people commute to work on their bicycles on the congested streets, I fear for their lives. Bicycle accidents and visits to emergency rooms are not uncommon. Beyond thinking about transportation risks, attending a public demonstration on behalf of climate change also has physical risks associated with it. Opposing

BOX 3.6. THE OVERRIDING COST/BENEFIT ANALYSIS

- Risks
  - Physical
  - Temporal
  - Social
  - Psychological
  - Functional
- Financial investments
- Perceived inequity
- System justification
- Conflicting goals, values, aspirations
- Tokenism

factions may be present. When considering whether to attend the demonstration, potential participants must factor in the physical risk of things careening out of control.

The *temporal risks* involve the time a person puts into learning about an issue, thinking about it, planning a course of action, perhaps actually even engaging in short or prolonged activity, and wondering whether the time spent was worth it. Whether it's thinking about your mode of transportation, being a vegetarian, or deciding whether to become involved in some environmental political action group, the thought about whether your time is being wasted is an important consideration.

Social risks concern the judgments that others may make about what we say and the actions we engage in. As mentioned previously, some people view environmentalists as "watermelons" or "eco-Nazis." A person espousing environmental concerns may be viewed as gullible and misguided for having fallen for the political left's "green scare." Losing one's standing in a family group or a community can be a major risk.

Psychological risks concern the risk of losing self-confidence or self-esteem after being criticized by others. People at times make those they disagree with feel small.

Functional risks involve whether, for instance, a new green technology will work. Purchasing a new green technology may involve a substantial cost. If a person is concerned, for instance, about the longevity of an electric battery in a car he's thinking of buying, and about the cost of replacing it, he may have second thoughts about taking the risk of making this purchase.

FINANCIAL INVESTMENTS People's previous purchase of a less eco-friendly product may also make them resistant to taking a financial loss by giving it up. For instance, if my daughter were to buy a new compact SUV, and then changed her mind about owning it immediately after taking it off the car lot, she'd never get all her money back. Disposing of it would be a bad financial decision, and the cost involved might make her resistant to changing vehicles again. Similarly, someone may be heavily invested in a stock that has been performing well. He might be hesitant to give up this stock for a more eco-friendly company if the latter company isn't as profitable. After all, his livelihood may depend on the earnings he receives from this company, and giving it up might be seen as adversely affecting his lifestyle.

PERCEIVED INEQUITY As I observed earlier when considering right-wing adherents' opposition to climate change policy, when people think that enacting proenvironmental laws or regulations might adversely affect the economy of the United States relative to other nations, they might resist supporting these laws or regulations. Seeing that other nations would gain an advantage at a cost to the American economy may prompt people to want to avoid suffering this cost.

SYSTEM JUSTIFICATION The desire to defend the status quo, especially when a person is benefiting from it, is another cost to consider. When individuals are experiencing a comfortable lifestyle, they may ask, "Why rock the boat?" They may see change as leading to less comfort and personal satisfaction, and this may persuade them to oppose others who are trying to change how things operate. As we saw earlier, this cost is a major concern among right-wing adherents. In combination with perceived inequity, this concern also points out that, when calculating costs and benefits, people may find certain costs and benefits more salient.

CONFLICTING GOALS, VALUES, ASPIRATIONS If you have led your life embracing a certain set of values and have gained meaning and purpose from trying to live in accordance with these values, being asked to change what has undergirded your existence may seem incomprehensible. After all, what would your life mean if you rejected what you once held as true and good? Would it all have been for naught? This is a lot to ask of a person, especially one whose status derives from meeting the goals associated with the mastery-oriented individualistic worldview.

TOKENISM Of course, a person could perform a minimal proenvironmental act, claim a favored identity as an environmentalist, and yet not actually change her behavior. In this way, she might gain social acceptance while incurring minimal, if any, cost. Helping, then, may entail greater or lesser acts, and people may encounter greater or lesser costs as a result. Some acts may be designed primarily for social display and social acceptance.

FINAL THOUGHTS: HOW CAN WE OVERCOME THE OBSTACLES TO HELPING?

Being Realistic about the Opposition and Barriers

Seeing the wide variety of barriers before us is important when considering how to overcome them. That some people benefit from the status quo and are resistant to change is a fact. That certain powerful interest groups do not want people to see clearly is also a fact. We need to be aware that people who benefit from the fossil fuel industry, for example, have a vested interest in maintaining a way of life dependent upon their products (see Mayer 2017). To see clearly is to understand that this opposition will not be swayed by facts and figures. To see clearly is to notice that some powerful others may wish to obscure our sight, make the emergency we face seem ambiguous, and stall change.

Equally important are the many other obstacles that can impede proenvironmental behavior. In this chapter I've pointed out that we are perceptually ill equipped to notice many of the changes in our world. And when coming to understand our world in terms of cause-effect relationships, it is difficult for us to see how small actions can lead to large outcomes, and how actions taken now can have impacts at a much later time and in a distant place. Moreover, we simply may not be psychologically prepared to face threats of extreme magnitude. These limitations hinder our ability to see and take action against the environmental challenges we face. In many ways we are blind to how we negatively affect the natural world. Becoming cognizant of our limitations may help us overcome them.

While there are many obstacles, there are also many practical things we can do to promote proenvironmental actions. For instance, time-lapse photography has been used to show people the effects of climate change and make them realize the emergency before us. The movie *Chasing Ice* is an excellent example of this technique, illustrating the extent to which glaciers have shrunk from the impact of a warming planet. Providing examples of effective proenvironmental actions can be helpful. Involving people in collective action can also help keep them from feeling individually overwhelmed.

As for thinking about how to design more effective environmental messages that motivate people to help, keep in mind that a message should never be thought of as stronger or weaker in and of itself. A message must always be considered in relation to the audience who will view it. If the message is aimed at liberals, then highlighting the negative impact that climate change will have on the most vulnerable—a message that takes into account their moral concerns—may be very effective. The same message may fail if delivered to conservatives, because they tend to embrace different moral concerns. When tailoring a message to conservatives, it might be more effective to state how climate change may harm the nation (i.e., their in-group) by increasing societal instability. Conservatives may also be more receptive to an appeal arguing that green technology will increase jobs and grow the economy. Unless the mitigative actions presented in a message can be portrayed as part of the system (Feygina, Jost, and Goldsmith 2010), those who benefit from the system may see change as a major cost.

De-emphasizing environmental issues in the message, and framing it instead as a cost-saving message or as a health message, is another approach to consider. For example, instead of arguing that people should use fuel-efficient cars in order to combat climate change, an argument could emphasize how the purchase of this car will save people money in the long run. Instead of creating a message criticizing coal on the basis of CO_2 emissions, an equally factual message could emphasize the negative impact that coal has on health, clean air, and clean water (i.e., moral concerns pertaining to purity).

Aside from the content of the message, the source of the message (i.e., who is delivering it) is another important factor to consider. Given that conservatives don't trust environmentalists or intellectuals, consider presenting a message from a traditionally conservative group or person. For instance, the Pentagon has produced a report delineating how climate change poses a threat to national security. Also consider referencing the pope's message on climate, or that of a Republican politician who has argued for the importance of combating climate change, as the source of the message.

The Great Transition

The major obstacle we need to overcome, however, is the U.S. worldview that distorts our sight. As both a cause of climate change and a barrier preventing people from addressing it, this mastery-oriented individualistic worldview is extremely problematic. Characterized by analytic thought, belief in a just world, and notions of progress, this worldview promotes a sense of superiority and has led us down a rosy path. But more and more the thorns are taking a toll.

We must also consider the true nature of the challenge before us. On the one hand, we can view environmental threats associated with climate change as the challenge we need to face. On the other hand, given that this challenge has largely been produced by people with a distorted view of reality, another way to construe the challenge is in terms of changing this worldview. Our view of ourselves as islands in an unlimited sea of resources is a distortion that has led to our present predicament: our view of material wealth and ever-greater consumption as the path to ever-greater happiness has led us to potential disaster. Fundamentally, in the depths of our minds, where our cultural lens frames our vision of the world, we need to change that vision. We need to see the view of separateness for what it is: an illusion. We need to recognize that, together, we are woven into a fabric of family, communities, and the natural world that extends into the future. We need to see that interconnectedness is the fundamental source of our happiness and future well-being.

A worldview that accurately depicts our lives as inextricably interwoven with the lives of others and the natural world will invariably lead us to notice, care, and act on behalf of others and the natural world. In other words, many of the barriers to helping depicted in this chapter arise from a cultural lens that distorts our vision and can be overcome by changing the lens through which we view our world. I articulate this alternative worldview in greater detail in the following chapters.

I hope that in the future, if a young woman cries for help, people will respond. Similarly, by transitioning to a worldview where people feel more connected not only to one another but also to the natural world, we can anticipate that people will be more likely to care for the distressed natural world we live in. As David Orr (2007) states, however, we should not passively, wishfully, hope for change but, rather, should recognize hope as a verb entailing action. To bring about the future we want, we must actively create it, by rolling up our sleeves and working for it.

CHAPTER FOUR

The Great Transition

From Separateness to Interconnectedness

> We were eating lunch on a high rimrock, at the foot of which a turbulent river elbowed its way. We saw what we thought was a doe fording the torrent, her breast awash in white water. When she climbed the bank toward us and shook out her tail, we realized our error: it was a wolf. A half-dozen others, evidently grown pups, sprang from the willows and all joined in a welcoming melee of wagging tails and playful maulings. What was literally a pile of wolves writhed and tumbled in the center of an open flat at the foot of our rimrock. . . . In a second we were pumping lead into the pack, but with more excitement than accuracy: how to aim a steep downhill shot is always confusing. When our rifles were empty, the old wolf was down, and a pup was dragging a leg into impassable slide-rocks. We reached the old wolf in time to watch a fierce green fire dying in her eyes. I realized then, and have known ever since, that there was something new to me in those eyes—something known only to her and to the mountain.
>
> Aldo Leopold, *Thinking Like a Mountain*

The Great Transition. What does it refer to? A change to a way of seeing the world that places humans in a harmonious relationship with nature. A change from the United States' worldview that emphasizes separateness, hierarchy, and materialism, to a new value-orientation that emphasizes connectedness, egalitarianism, and community. It entails more than simply increasing environmentally sustainable practices. Research has demonstrated that this new orientation produces personal well-being, enhanced personal and collective efficacy, greater cooperation, stronger communities, enhanced problem-solving ability, and creativity. At the end of the chapter

I'll consider the implications of how this change might bring about harmonious functioning *within* human systems as well. After all, how can there be harmony between human and environmental systems if human systems are fraught with dissension, conflict, and war? I'll also consider implications of how this change might affect our lives on Eaarth.

Let me start by making one important point. In the prologue, I stated that science enables us to measure aspects of our world and critique our worldviews. In this capacity, science can help us see more clearly, chart a different course of action, and remap our relationship to the natural world. After all, if I am asking you to transition to an alternative worldview, it is important for you to have a scientifically grounded and nuanced view of this alternative worldview. The items that measure this alternative worldview help articulate its meaning. Measures of this alternative worldview empirically relate to other measures (e.g., proenvironmental behavior and well-being), and these relationships support my claim that benefits derive from the transition to this worldview. Looking at the stars at night and recognizing the cluster of stars that make up Ursa Major (the greater she-bear), and seeing how these clustered stars connect and give rise to this image, is like seeing how the cluster of discrete items on a scale gives rise to the image of an alternative worldview. And in the same way that Ursa Major can aid a traveler by pointing to the north star, this alternative worldview not only helps us see better but also helps us set our course.

Let's begin, however, by considering the personal transition that Aldo Leopold experienced in his life. Leopold is considered one of the great environmentalists of the modern age. The first steps toward his shift in sight occurred on that Arizona mountainside. Up to the encounter with the old wolf, he accepted the prevailing view that the only good wolf was a dead wolf. But then he saw the "fierce green fire dying in her eyes," and everything changed for him. He could no longer think of the wolf as a type of vermin that needed to be destroyed. He experienced something "new" in those eyes. What this was I can only conjecture. In light of his writings, however, I think it's fair to say that he saw wisdom and intelligence in the eyes of that wolf. He experienced a connection with the wolf that led to empathy and compassion. He realized the commonality, the oneness, the community, and the kinship that we share with nature. After this encounter, the human/nature split vanished for him.

THE LAND ETHIC WORLDVIEW

Leopold makes his point more explicitly in his chapter on the land ethic (Leopold 1949). He envisions the land ethic as the alternative worldview that nurtures greater harmony between human and environmental systems. In this worldview, the "I" and "other" distinction dissolves, replaced by the "we" of connectedness and interrelatedness between humans and nature. He states, "The land ethic simply enlarges the boundaries of the community to include soils, waters, plants, and animals, or collectively: the land" (p. 239). This ethic calls for humans to develop a different view of their relationship to the natural world. In this worldview, there is not a mastery relationship between humans and nature. Instead, there is an egalitarian, kinship relationship. This is an ecological, or biospheric, worldview. The interactions between living organisms, the communities they belong to, and the nonliving, or abiotic, world are all considered as a whole.

The land ethic worldview expands and strengthens our sense of community. Instead of seeing self and others as isolated islands, people now see themselves as connected with one another and with nature (refer to figure 4.1). These bonds create a sense of community and interdependence, not separateness. Additionally, these ties create a sense of oneness. The land ethic worldview leads people to see themselves as, not greater or lesser than other elements, but similar in size. In other words, as Leopold states, the land ethic worldview leads people to see themselves as plain and simple members of the natural world.

As a consequence of seeing oneself as equal to and connected to nature, the land ethic calls for our scope of justice to be extended beyond human affairs and to include the rights of nature. The implication that our ethic can be extended to nature is beautifully captured in the opening paragraphs of Leopold's chapter on the land ethic from his seminal book *The Sand County Almanac* (1949, p. 237).

> When god-like Odysseus returned from the wars of Troy, he hanged all on one rope a dozen slave-girls of his household, whom he suspected of misbehavior during his absence. This hanging involved no question of propriety. The girls were property. The disposal of property was then, as now, a matter of expediency, not of right and wrong. Concepts of right and wrong were not lacking from Odysseus' Greece: witness the fidelity of his wife through the

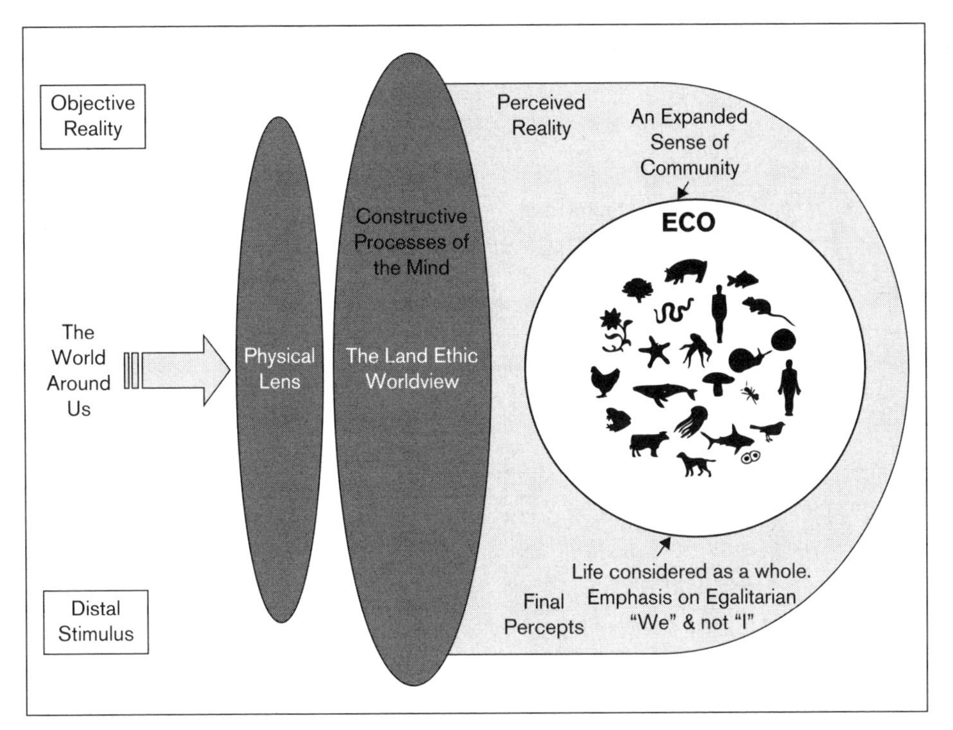

Fig 4.1. The land ethic worldview. This ecological worldview is characterized not by a mastery-and-hierarchical relationship but by an egalitarian kinship relationship between humans and nature.

> long years before at last his black-prowed galleys clove the wine-dark seas for home. The ethical structure of that day covered wives, but had not yet been extended to human chattels. During the three thousand years which have since elapsed, ethical criteria have been extended to many fields of conduct, with corresponding shrinkages in those judged by expediency only.

For Leopold on that Arizona mountainside, it was expedient to rid the area of wolves. Wolves were predators of deer. It served Leopold's purpose, as a deer hunter, to eliminate the wolves so that the deer population would flourish. But with the intimacy, empathy, compassion, and respect gained from seeing the green fire die in the old wolf's eyes, Leopold made a connection between himself and the wolf. His sense of community and ethical concerns were extended beyond human communities to include the collective natural community (i.e., "the land").

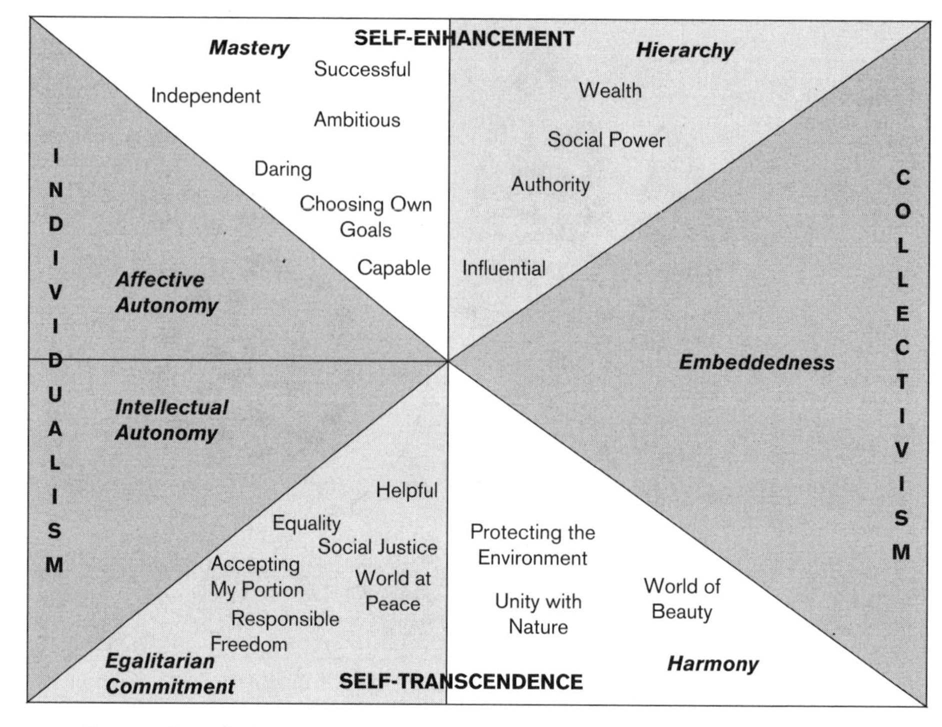

Fig 4.2. Adapted Schwartz Value Model. S. H. Schwartz, "Beyond Individualism/Collectivism: New Cultural Dimensions of Values," in *Individualism and Collectivism: Theory, Method, and Applications*, edited by U. Kim, H. C. Triandis, C. Kagitcibasi, S.-C. Choi, and G. Yoon (Thousand Oaks, CA: Sage, 1994), adapted from figure 7.1.

Relating Leopold's land ethic to Schwartz's value model (refer to figure 4.2), the land ethic represents a shift from a *mastery*-oriented individualistic value orientation to an *egalitarian* individualism or *harmony* value orientation. It illustrates a shift from a self-enhancement or egoistic orientation to a self-transcendent orientation, or what others refer to as an ecological or biospheric orientation. Humans and nature are placed in a more equal relationship. The hierarchy that enthrones humans as the life-form with dominion over the world is replaced by a stronger sense of oneness, connectedness, and caring for the natural world.

At the end of the opening passage of Leopold's story of the old wolf, he makes reference to "something known only to her and to the mountain." To what does this cryptic statement refer? For Leopold, it involved a different way of thinking that captured a holistic relationship, or systems-thinking orientation. He came to realize that, while the deer feared the wolves, in some sense the mountain feared the deer, for a proliferation of deer would strip the mountain of its foliage. He saw connections and interrelationships not only in terms of feeling closer to members of the enlarged community but also with respect to how they affected one another. Instead of seeing the elements (the mountain, the deer, the wolf, trees, bushes, and soil) as isolated from one another, he began to see the big picture. As Leopold put it,

> Since then I have lived to see state after state extirpate its wolves. I have watched the face of many a newly wolfless mountain, and seen the south-facing slopes wrinkle with a maze of new deer trails. I have seen every edible bush and seedling browsed, first to anaemic desuetude, and then to death. I have seen every edible tree defoliated to the height of a saddlehorn. Such a mountain looks as if someone had given God a new pruning shears, and forbidden Him all other exercise. In the end the starved bones of the hoped-for deer herd, dead of its own too-much, bleach with the bones of the dead sage, or molder under the high-lined junipers.
>
> I now suspect that just as a deer herd lives in mortal fear of its wolves, so does a mountain live in mortal fear of its deer. And perhaps with better cause, for while a buck pulled down by wolves can be replaced in two or three years, a range pulled down by too many deer may fail of replacement in as many decades. So also with cows. The cowman who cleans his range of wolves does not realize that he is taking over the wolf's job of trimming the herd to fit the range. He has not learned to think like a mountain. Hence we have dustbowls, and rivers washing the future into the sea. (1949, pp. 139–140)

Feeling a greater sense of connection to the natural world involves seeing more accurately how our actions affect the natural world. It helps us understand how changes in one area of a system reverberate through the entire system. It is the realization that killing wolves led to more deer, more deer to less foliage on a mountain, less foliage on a mountain to topsoil loss and rivers washing this precious resource to the sea. This transition from separateness to connection and from analytic to more holistic thinking can help us see and change course, so that our own "too-much" in the form of overpopulation and overconsumption doesn't lead to our bleached bones on a path of human folly.

Measuring the Land Ethic Worldview: The Connectedness to Nature Scale

My colleague Cindy Frantz and I developed the Connectedness to Nature Scale to measure the extent to which a person has adopted a land ethic worldview (Mayer and Frantz 2004). Here, I simply refer to this scale as the Land Ethic Scale. The items on this scale emphasize experiencing self as being connected to nature, feeling in community with nature, feeling kinship with animals and plants, and experiencing self as a plain and simple member of the natural world. The scale is presented in box 4.1.

This scale has good psychometric properties, meaning that, as a measure, it has desirable characteristics. This is an important consideration because basing an argument on the findings associated with a scale that isn't measuring what it intends to measure (i.e., a scale that isn't valid) or isn't reliable would provide a weak argument. Let me point out a few interesting characteristics of the scale. First, although the individual items that comprise the scale tap into different thoughts and feelings, statistically these items are strongly associated with one another. This finding indicates that the scale has high internal reliability, and that it represents a cohesive worldview. Second, people who feel more or less connected to nature at one time period have been shown to remain so across time (i.e., the scale is said to have high temporal reliability). This finding argues that feeling more or less connected to nature is not an ephemeral experience but can be thought of as a relatively stable worldview that characterizes how people view their relationship to the natural world. Although this worldview is relatively stable, it can fluctuate depending upon the immediate setting that a person is in. Time spent in a natural setting can momentarily heighten people's scores on this scale in comparison to when they are in an urban setting. Lastly, research demonstrates that this scale is related to things it should be related to and is independent of things it should not be related to. These findings argue for the validity of the scale, indicating that it is measuring what it intends to measure (i.e., that the scale has high predictive, convergent, and discriminant validity).

The advantage of developing a scale to measure the land ethic is that it enables us to empirically investigate whether this worldview suffers from the same shortcomings as the mastery-oriented individualistic worldview. Instead of relying on verbal arguments for whether this worldview does or

BOX 4.1. THE CONNECTEDNESS TO NATURE SCALE (THE LAND ETHIC SCALE; MAYER AND FRANTZ 2004)

1. I often feel a sense of oneness with the natural world around me.
2. I think of the natural world as a community to which I belong.
3. I recognize and appreciate the intelligence of other living organisms.
4. I often feel disconnected from nature. (Reverse worded.)
5. When I think of my life, I imagine myself to be part of a larger, cyclical process of living.
6. I often feel kinship with animals and plants.
7. I feel as though I belong to the earth as equally as it belongs to me.
8. I have a deep understanding of how my actions affect the natural world.
9. I often feel part of the web of life.
10. I feel that all inhabitants of the earth, human and nonhuman, share a common "life force."
11. Just as a tree can be part of a forest, I feel embedded within the broader natural world.
12. When I think of my place on earth, I consider myself a top member of a hierarchy that exists in nature. (Reverse worded.)
13. I often feel like only a small part of the natural world around me, and that I am no more important than the grass on the ground or the birds in the trees.
14. My personal welfare is independent of the welfare of the natural world. (Reverse worded.)

doesn't do this or that, we can actually test ideas. Through scientific research, we can establish what this worldview is and isn't associated with.

Other Related Measures of Connectedness to Nature

The Land Ethic Scale is the only scale to measure individuals' experience of feeling an egalitarian relationship to the natural world, communality with nature, and a sense of kinship with nature. That said, there are several other related connectedness-to-nature scales. These scales, for different theoretical reasons, highlight the importance of feeling a sense of oneness with nature. Let me briefly introduce these scales and comment on each of them.

Two scales are directly concerned with measuring individuals' feelings of love of nature (Kals, Schumacher, and Montada 1999; Perkins 2010). Sample items from each of these scales are presented below (refer to box 4.2).

The Emotional Affinity toward Nature Scale, created by Kals, Schumacher, and Montada (1999), is based on the theoretical reasoning that people have a need to affiliate with nature. This reasoning is based on Fromm's (1964) concept of biophilia, which refers to a growth-oriented aspect of people related to love of life and of everything that lives and grows. It is also based on Wilson and Kellert's biophilia hypothesis (Wilson 1984; Kellert and Wilson 1993). This hypothesis argues that because the human species has evolved in natural settings, it has developed a biologically based need to affiliate with and feel connected to the broader natural world. In developing their scale, Kals, Schumacher, and Montada saw "feeling good, free, safe in nature, and feeling oneness with nature," as other characteristics "closely related to love of nature" (1999, p. 182).

In contrast, the Love and Care for Nature Scale was developed by Perkins (2010) to tap into "(1) feelings of awe, wonder, and interest in nature, which are sustained emotions said to evoke feelings of care; (2) feelings of love, emotional closeness and interconnectedness with nature, including a spiritual aspect somewhat neglected in the psychological literature; and (3) feelings of care, responsibility and commitment to protect nature." The scales devised by Kals, Schumacher, and Montada (1999) and Perkins (2010) both emphasize that caring for the natural world involves more than a cold, cognitive appraisal. From these authors' perspective, we help both people and nature because of the emotional bonds that are present.

Both of these scales derive from traditions very different than those of the Land Ethic Scale. Nonetheless, Leopold (1949) does state, "We abuse land because we regard it as a commodity belonging to us. When we see land as a community to which we belong, we may begin to use it with love and respect." These scales clearly do not portray nature as a commodity. Moreover, while the land ethic's emphasis on communality isn't explicitly stated, these scales do emphasize the importance of feeling a close connection and

BOX 4.2. EMOTIONALLY BASED CONNECTEDNESS TO NATURE SCALES

Emotional Affinity toward Nature Scale (Kals, Schumacher, and Montada 1999)

Sample items:

1. If I spend time in nature today, I feel a deep feeling of love toward nature.
2. When I spend time in nature I feel free and easy.
3. I have the feeling I can live my life to the full in nature.
4. When surrounded by nature I get calmer and I feel at home.
5. Whenever I spend time in nature I do experience a close connection to it.
6. Sometimes when I feel unhappy I find solace in nature.

Love and Care for Nature Scale (Perkins 2010)

Sample items:

1. When I am close to nature, I feel a sense of oneness with nature.
2. I feel content and somehow at home when I am in unspoilt nature.
3. I often feel emotionally close to nature.
4. I feel spiritually bound to the rest of nature.
5. I often feel a sense of awe and wonder when I am in unspoilt nature.
6. I often feel a strong sense of care towards the natural environment.
7. When in nature, I feel emotionally close to nature.

oneness with nature. What I find so interesting about these scales is that they suggest the "love and respect" Leopold talks about may derive not only from communality with nature but also from the comfort and freedom one can experience in nature, and the awe and wonder that nature inspires. In other words, different paths may lead to love, respect, and by implication, caring for the natural world.

It shouldn't be surprising, then, that the Land Ethic Scale is highly correlated with both the Emotional Affinity Scale and the Love and Care for Nature Scale. That the Land Ethic Scale is so strongly associated with these scales extends the meaning of the Land Ethic Scale. If we think of the Land Ethic

Scale as a constellation of fourteen items that cluster together, we can see this constellation expand as we find other items closely associated with it. Feelings of love toward nature, caring for the natural world, feeling free and easy when in nature, experiencing awe and wonder, and feeling able to live life more fully when in nature are all part of this alternative worldview. People viewing the world through the land-ethic cultural lens do not feel alone and isolated, as someone with a mastery-oriented individualistic orientation might feel. They are not walking through a lifeless world but experiencing the awe and wonder of nature. They experience a sense of belonging to this natural community, a sense of comfort and of being at home in nature, and a sense of support that derives from a close emotional relationship with the world around them. A point I will elaborate on later in the chapter is that this worldview is associated with not only greater caring for and protection of nature but also a greater vitality and sense of well-being. Underlining the transition from the self-enhancement orientation to a more self-transcendent orientation, a spirituality component, too, emerges as part of the land ethic constellation.

Davis, Green, and Reed's (2009) Commitment to the Environment Scale approaches the concept of experiencing a sense of connectedness to nature from yet a different theoretical direction (refer to box 4.3). The authors base their work on interpersonal relationships theory. Extending this theory to the natural world, their work emphasizes relationship commitment derived from interdependence theory. They think of the natural world as a potential relationship partner and consider the extent to which an individual is dependent upon the partner to satisfy needs.

> To the degree that individuals perceive that they are dependent on the natural environment for their own well-being (e.g., they derive satisfaction from scuba diving), they will experience a corresponding level of commitment to the environment (e.g., they will be interested in maintaining the well-being of coral reefs over time). From the perspective of interdependence theory, such interest in the well-being of the environment has a structural explanation (i.e., dependence on the environment for the unique gratification of needs) and not a connectedness or closeness explanation (e.g., nature is part of my self-concept). (pp. 174–175)

In contrast, Clayton's (2003) Environmental Identity Scale (refer to box 4.3) derives from research on self-identity and individuals' self-concepts.

BOX 4.3 COMMITMENT TO THE ENVIRONMENT SCALE (DAVIS, GREEN, AND REED 2009)

Sample items:

1. I am interested in strengthening my connection to the environment in the future.
2. I feel strongly linked to the environment.
3. When I make plans for myself, I take into account how my decisions may affect the environment.
4. I believe that the well-being of the natural environment can affect my own well-being.
5. I feel committed to keeping the best interests of the environment in mind.

Environmental Identity Scale (Clayton 2003)

Sample items:

1. In general, being part of the natural world is an important part of my self-image.
2. I spend a lot of time in natural settings (woods, mountains, desert, lakes, ocean).
3. I think of myself as part of nature, not separate from it.
4. When I am upset or stressed, I can feel better by spending some time outdoors.
5. I feel that I receive spiritual sustenance from experiences in nature.
6. Engaging in environmental behaviors is important to me.
7. If I had enough time or money, I would certainly devote some of it to working for environmental causes.
8. My own interests usually seem to coincide with the position advocated by environmentalists.

According to Clayton, in the same way that a person can feel connected to a group and maintain a collective identity (e.g., an ethnic identity), a person can also feel connected to nature and develop an environmental identity. Nature can be part of a person's sense of who he is. When a person feels connected to nature, then, he no longer views nature as the "other" or as lifeless. Rather, when people experience a common essence and strong ties, the human/nature split vanishes.

Both the Commitment to the Environment Scale and the Environmental Identity Scale are also highly correlated with the Land Ethic Scale and expand the meaning of the Land Ethic Scale even further. Thinking of nature as a relationship partner leads people to consider the well-being of nature as essential to their own personal well-being. This relationship emphasis coincides very closely with the land-ethic notion of communality with nature. It persuades people to commit themselves to protecting the well-being of nature and to intentionally think about how their plans affect the environment. They feel committed to keeping the best interests of the environment in mind. And, as emphasized in Clayton's scale, when people no longer think of nature as the "other," they engage in proenvironmental behaviors (the last three sample items from the Environmental Identity Scale).

Clayton's scale also brings up several other interesting points, which pick up on themes presented in the Emotional Affinity toward Nature Scale: nature provides solace and is a place where we can go to relieve stress. The spiritual dimension, too, is brought up in this scale. Overall, while the Land Ethic Scale focuses on the idea that a person can be an egalitarian member of a natural community, and on the kinship we share with nature, these other scales elaborate upon how this focus is tied to so many other feelings and experiences. The sense that we are alive in a world that is alive, that we share a common essence with nature, and that our well-being is tied to the well-being of nature is a worldview far removed from the mastery-oriented individualistic worldview.

Nisbet, Zelenski, and Murphy (2009) developed the Nature Relatedness Scale, composed of three subscales: Nature Relatedness—Self; Nature Relatedness—Perspective; and Nature Relatedness—Experience (refer to box 4.4). The Nature Relatedness—Self subscale is designed to tap into a person's thoughts and feelings about her personal connectedness to nature. The Nature Relatedness—Perspective subscale represents "an external nature-related worldview, a sense of agency concerning individual human actions and their impact on all living things" (p. 723). As for the Nature Relatedness—Experience subscale, it was intended to tap into a person's "physical familiarity with the natural world" and "the level of comfort with

BOX 4.4 NATURE RELATEDNESS SCALE (NISBET, ZELENSKI, AND MURPHY 2009)

Sample items:

Nature Relatedness—Self

1. My connection to nature and the environment is part of my spirituality.
2. I feel very connected to all living things and the earth.
3. My relationship to nature is an important part of who I am.
4. I always think about how my actions affect the environment.
5. I think a lot about the suffering of animals.
6. Even in the middle of the city, I notice nature around me.

Nature Relatedness—Perspective (all items reverse worded)

1. Humans have the right to use natural resources any way we want.
2. Animals, birds and plants have fewer rights than humans.
3. Some species are just meant to die out or become extinct.
4. Nothing I do will change problems in other places on the planet.
5. Conservation is unnecessary because nature is strong enough to recover from any human impact.

Nature Relatedness—Experience

1. My ideal vacation spot would be a remote, wilderness area.
2. I enjoy being outdoors even in unpleasant weather.
3. I don't often go out in nature. (Reverse worded.)
4. I enjoy digging in the earth and getting dirt on my hands.
5. I take notice of wildlife wherever I am.

and desire to be out in nature" (p. 725). What I find most interesting about this scale is that, as indicated by Perkin's Love and Care for Nature Scale and Clayton's Environmental Identity Scale, part of feeling connected to nature involves a sense of spirituality. Additionally, another item on the Nature Relatedness—Self Scale indicates that people who feel connected to nature are not indifferent to the suffering of the natural world (sample item 5). As you'll see, there is no distancing between self and nature here as there is in the mastery-oriented individualistic worldview.

The Nature Relatedness—Self Scale is highly correlated with the Land Ethic Scale, indicating once again a high degree of commonality between these scales. That the Nature Relatedness—Experience subscale is moderately correlated with the Land Ethic Scale speaks to the comfort and familiarity of nature, and the desire to be immersed in nature, which is associated with the land ethic worldview. That the Nature Relatedness—Perspective subscale shows a weak but significant correlation with the Land Ethic Scale suggests a point I elaborate upon later: that the land ethic worldview is associated with agency (i.e., self-efficacy) and the greater awareness some people have of their impact on the natural world.

Lastly, sample items from the scales by Dutcher, Finley, Luloff, and Johnson (2007) and Hedlund-de Witt, Boer, and Boersema (2014) are presented in Box 4.5. As you can see by examining these sample items, both scales tap into a person's sense of feeling connected to nature. However, like others that I've already commented on, neither scale asks questions about a person's feeling of an egalitarian relationship to the natural world, or feeling as if in a community with nature, or a person's sense of kinship with nature. I think it is interesting that Hedlund-de Witt, Boer, and Boersema's measure does include items (sample items 2 and 5) that, like the Nature Relatedness—Self Scale, also tap into a basic sense of strong concern for nature, and not indifference. Additionally, item 6, concerning respect for nature and our adjustment to nature, as opposed to forcing nature to adjust to us, captures a nondominating stance toward the natural world. As you might expect, the Connectedness with Nature Scale, by Dutcher, Finley, Luloff, and Johnson, is highly correlated with the Land Ethic Scale, and I suspect the Land Ethic Scale is highly correlated with the scale by Hedlund-de Witt, Boer, and Boersema, as well.

Final Comments on Scales

The Land Ethic Scale is composed of a cluster of items that form a constellation of meaning. These other related connectedness-to-nature scales expand this constellation, extending the meaning of communality with nature to include a host of other meanings. I find it amazing that researchers from such different theoretical directions (land ethic, biophilia, interpersonal relationships, self-concept, and identity) have all ended up at the

BOX 4.5 CONNECTEDNESS WITH NATURE SCALE (DUTCHER, FINLEY, LULOFF, AND JOHNSON 2007)

Sample items:

1. I see myself as part of a larger whole in which everything is connected by a common essence.
2. I feel a sense of oneness with nature.
3. The world is not merely around us but within us.
4. I never feel a personal bond with things in my natural surroundings, like trees, a stream, wildlife, or the view on the horizon. (Reverse worded.)

Note: This scale also includes the Nature in Self Scale (i.e., the INS).

Connectedness to Nature Scale (Hedlund-de Witt, Boer, and Boersema 2014)

Sample items:

1. I have a deep feeling of connectedness to nature.
2. It hurts me to see nature destroyed.
3. Things that I enjoy, but are bad for the environment, I want to keep doing. (Reverse worded.)
4. I aspire to a conscious and more natural lifestyle.
5. I don't care so much that species are becoming extinct. (Reverse worded.)
6. The relationship between human beings and nature should be one of respect, adjustment and attunement.

same place, emphasizing connectedness, closeness, and oneness with nature as the path to proenvironmental behavior and well-being. A core set of elements ties these scales together (cf. Tam 2013; refer to box 4.1). That caring for the natural world involves dissolving the distance between self and nature seems like the perfect antidote to the mastery-oriented individualistic worldview that nurtures separateness and indifference.

The presence of different scales measuring connectedness to nature in different ways is important for several reasons. First, although the scales

TABLE 4.1
Summary of Correlations between the Land Ethic Scale and Other Related Scales

	Land Ethic Scale
Emotional Affinity toward Nature Scale	.71
Love and Care for Nature Scale	.79
Commitment to the Environment Scale	.81
Environmental Identity Scale	.77
Nature Relatedness Scale	
Nature Relatedness—Self	.72
Nature Relatedness—Perspective	.23
Nature Relatedness—Experience	.44
Connectedness with Nature Scale	.72

share a basic core, they are unique as well. They ask similar but also different questions. For instance, that the Land Ethic Scale is strongly correlated with scales that include questions about concern for the harm being done to the natural world makes me think that people scoring higher on the Land Ethic Scale also share this concern (refer to sample item 5 from Nature Relatedness—Self subscale, and sample items 2 and 5 from the Connectedness to Nature Scale by Hedlund-de Witt, Boer, and Boersema). That the Land Ethic Scale is strongly associated with scales that include a spirituality component prompts me to wonder about this association (refer to sample item 4 from Love and Care Scale, sample item 5 from Environmental Identity Scale, and sample item 1 from Nature Relatedness—Self Scale).

A second and perhaps more important point, though, is that if the findings in this area were solely based on one scale, there would always be a question about whether the findings might be due to some idiosyncratic aspect of that particular measure. However, when you have multiple measures of connectedness to nature revealing the same things, you have added confidence in the results. Moreover, that the findings of different researchers—conducting their work with different participant populations, and using different measures of proenvironmental behavior and well-being—all converge and conceptually replicate one another speaks to the validity, robustness, and generality of these findings. Taken together, it is hard to dismiss the findings as owing to an idiosyncratic and imperfect characteristic of any given connectedness scale or outcome measure or to the use of only a specific subset of the population.

Lastly, it is also important to realize that a variety of research methodologies has been used. While much of the research is correlational in nature, experimental and longitudinal work is also present. Regarding the correlational work, I emphasize the strength of the associations. They are remarkably high. Additionally, regarding the correlational work, experimental control may be absent, but statistical controls are often present. For instance, as I mentioned in chapter 1, Frantz and I (Mayer and Frantz 2004) statistically controlled for participants' New Environmental Paradigm scores when investigating the impact of Land Ethic Scale scores on lifestyle habits, perspective-taking, and proenvironmental behavior. In other words, when examining the association of Land Ethic Scale scores with each of these different outcome variables, we eliminated the alternative explanation that the associations we observed were due to a third variable (i.e., participants' New Environmental Paradigm scores). This is just one statistical technique that researchers use.

Fearing that a focus on statistical analyses and the strengths and weaknesses of different methodological approaches will disrupt the narrative flow, when presenting each study here I do not emphasize these more nuanced issues. However, the important point to keep in mind is that different kinds of experimental control and statistical control undergird much of this research. Thus, in much the same way that multiple scales contribute to stronger research conclusions, the wide variety of methodologies and statistical analyses found in this research also contributes to a stronger research base from which to draw conclusions.

Since the Land Ethic Scale most clearly taps into Leopold's land ethic worldview, this chapter highlights research that has used this scale. However, given the close associations between this scale and these other related connectedness-to-nature measures, I present findings associated with them as well, as a way to corroborate the Land Ethic Scale findings. Now that we've briefly considered these different measures, let's turn to research investigating how this worldview is related to a number of important issues.

TABLE 4.2

Correlations between the Land Ethic Scale and Other Related Scales with Schwartz Value Dimensions

	Land Ethic	Emotional Affinity	Commitment to the Environment	Environmental Identity	Nature Relatedness
Self-enhancement	.07	.14	.13	.11	.10
Self-transcendence	.45	.52	.51	.52	.49

SOURCE: Tam 2013.

THE LAND ETHIC, SELF-ENHANCEMENT, AND SELF-TRANSCENDENCE

I've argued that the land ethic worldview represents a shift from a mastery-oriented individualistic value orientation to an egalitarian-oriented individualistic or harmony value orientation. Stated differently, it represents a shift from a self-enhancement, or egoistic, orientation to a self-transcendent—or what others call an ecological or biospheric orientation—in which humans and nature are placed in a more equal relationship.

To test this hypothesis, we can examine whether the Land Ethic Scale and other related connectedness-to-nature measures are more strongly associated to a measure of Schwartz's self-transcendence dimension than to a measure of his self-enhancement dimension. Tam (2013) tested this hypothesis by having participants fill out the Schwartz Value Questionnaire. Supporting the argument that the land ethic worldview is more strongly associated with self-transcendence than with self-enhancement, the Land Ethic Scale and each of the other related connectedness-to-nature measures were found to be weakly associated or not associated with the self-enhancement dimension, and far more strongly associated with the self-transcendence dimension, as you can see in table 4.2. This finding helps establish the land ethic worldview as in fact representing a shift toward a value orientation where humans and nature exist in greater harmony with one another.

In other research examining whether the Land Ethic Scale is associated with an egoistic orientation (mastery) or a greater caring for others and the natural world (egalitarian or harmony orientation), Frantz and I (Mayer and Frantz 2004) investigated the relationship between the Land Ethic Scale and a general value scale measuring endorsement of egoistic (mastery) or egalitarian/biospheric values. In two studies, we framed our questions in terms of the extent to which a person is concerned about environmental problems owing to their consequences for that individual's personal well-being (egoistic value orientation), their consequences for family, friends, and community (social-altruistic value orientation), or their consequences for plants, trees, and animals (a biospheric value orientation), and we found the Land Ethic Scale to be most strongly associated with the biospheric orientation and not related to an egoistic orientation. In other words, the land ethic is associated with the value of caring for the natural world, independent of egoistic concerns. This clearly reflects the self-transcendent orientation associated with the egalitarian and harmony value orientations.

Given that the Land Ethic Scale represents an egalitarian value base, it is also important to consider the relationship between egalitarianism and caring for the natural world. Several studies have examined this issue. For instance, Price, Walker, and Boschetti (2014) found strong support for the relationship between egalitarianism and environmentalism. Other researchers found that egalitarianism is associated with support for raising energy taxes, slowing industrial growth (Carlisle and Smith 2005), and support for climate change policies (Leiserowitz 2006). Lastly, Portinga, Steg, and Vlek (2002) found that egalitarianism is associated with support for energy efficiency, public transportation, and carpooling. By implication, then, given that the Land Ethic Scale is related to egalitarianism, and egalitarianism is related to caring for the natural world, we can see yet another link between the land ethic worldview and proenvironmental behavior.

THE LAND ETHIC AND DISTANCING EFFECTS

Recall that one major criticism of the U.S. worldview was that it distances individuals from nature. This distancing, it was argued, enabled people to harm nature with psychological impunity. Let's see how the Land Ethic Scale and the other related connectedness-to-nature measures are associated with a measure of self/nature distance. The Inclusion of Nature in Self Scale (Schultz 2001) is presented in figure 4.3.

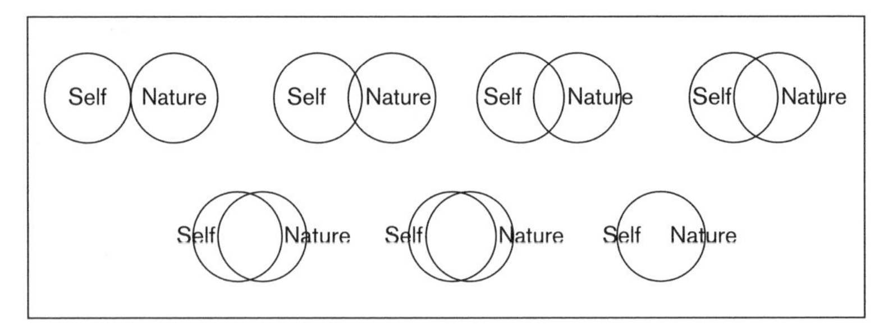

Fig 4.3. The Inclusion of Nature in Self Scale, from P. W. Schultz, "Assessing the Structure of Environmental Concern: Concern for the Self, Other People, and the Biosphere," *Journal of Environmental Psychology* 21:327–339.

TABLE 4.3

Correlations between the Land Ethic Scale and Related Scales with Distancing of Self from Nature

	Land Ethic	Emotional Affinity	Commitment to the Environment	Environmental Identity	Nature Relatedness
Distancing	.64	.59	.48	.46	.63

SOURCE: Tam 2013.

As this scale illustrates, self and nature are, at one extreme, independent of one another (i.e., they do not overlap at all). At this extreme, self and nature can be thought of as distanced from one another. At the other extreme, there is complete overlap between self and nature. At this extreme, there is no distance between self and nature. If the land ethic does lead to less distancing, then we should expect the Land Ethic Scale and the other related connectedness measures to be positively associated with the Inclusion of Nature in Self measure.

Referring to table 4.3, you can see that for each of the connectedness-to-nature measures, Tam (2013) did find a moderate-to-strong association with the Inclusion of Nature in Self measure. These findings indicate that the self/nature divide is diminished when people feel more connected to nature. Frantz and I (Mayer and Frantz 2004, study 5) also found that the Land Ethic Scale shows a moderate-to-strong correlation with this scale ($r = .55$), indicating that the individuals who hold the land ethic worldview see a "oneness" between self and nature and do not distance themselves from the natural world.

Further corroboration of this finding comes from Dutcher, Finley, Luloff, and Johnson (2007). Recall that the Inclusion of Nature in Self Scale is actually part of their broader Connectedness with Nature Scale. In statistically examining the goodness of their overall scale, they performed a factor analysis of each item from their Connectedness with Nature Scale with the Inclusion of Nature in Self Scale to see whether the two scales were each measuring something in common. The results of this analysis did in fact show a commonality between the items on their scale and the Inclusion of Nature in Self measure. Thus, examining the distancing phenomena in a different way and with a different measure, they observed the same findings. In other words, it seems clear that the land ethic worldview is associated with less distancing between the self and the natural world.

These findings may seem obvious to you. After all, doesn't connection mean lack of distance? In the Inclusion of Nature in Self Scale, however, even on the left side, which represents the greatest distance between self and nature, the self-circle and the nature-circle are still touching one another. In some minds, this might represent connection. What is so interesting about these findings is the observed overlap. Greater connectedness to nature is associated with a merging of self and nature. The two become indistinguishable from one another. As in the case of members of a family, it's not just that you may live near one another but, more importantly, the perceived similarity, a sharing of some common essence, that unites the members beyond a simple physical proximity.

THE ISSUE OF MATERIALISM

Another major criticism of the U.S. worldview is that it is associated with an emphasis on materialism, consumption, and a freedom to use nature in an instrumental fashion to achieve personal desires. This is the trampling-over-nature part of the mastery-oriented individualistic worldview that, in combination with distancing, produces such harm to the natural world.

Using their Connectedness to Nature Scale, Hedlund-de Witt, Boer, and Boersema (2014) found that connectedness to nature is negatively related to a focus on money. In other words, in comparison to people who scored lower on connectedness to nature, people who reported feeling more connected to nature indicated that earning a lot of money was less important to them; they reported being less likely to aspire to a luxurious and comfortable lifestyle, and less likely to see their quality of life as tied to how much money they spent. Frantz and I (Mayer and Frantz 2004) also found that the Land Ethic Scale is negatively associated with a measure of consumerism. Like Hedlund-de Witt, Boer, and Boersema, we also found that being more connected to nature was associated with placing less importance on buying things.

The land ethic worldview, then, counters the destructive aspects of the mastery-oriented individualistic worldview as it relates to the natural world. As presented in chapter 2, the United States' mastery-oriented individualistic worldview nurtures a combination of cultural striving that harms the natural world and a psychological distancing of individuals from nature that enables people to engage in this harmful behavior with psychological impunity. In contrast, the land ethic worldview leads people to feel closer to nature and less concerned with consumerism. More of the egalitarian and harmony values come into play.

THE LAND ETHIC, PERSPECTIVE-TAKING, AND ENVIRONMENTAL CONCERN

Does the "oneness" of self and nature that characterizes the land ethic worldview promote perspective-taking? As presented in the discussion of the Latane and Darley helping model in chapter 3, a key factor that affects helping behavior is the sense of "we," of connectedness and interrelatedness between the potential helper and whoever or whatever needs aid. The importance of a sense of "we" cannot be emphasized enough. This sense of connection and interrelatedness fosters perspective-taking. Feeling a sense of "we" enables a person to "walk in the shoes of the other" or, for Leopold, to "think like a mountain" and experience the "other's" dilemma as your own. As Roszak (1992) said years ago, when the human/nature split vanishes and nature is experienced as part of who you are, then nature's pain becomes your own personal pain, and your environmental concern and motivation to help increases.

It is fair to say that much of the literature on connectedness to nature was motivated by researchers' desire to address the environmental threats before us. As Leopold experienced on that Arizona mountain, feeling connected to the wolf changed everything for him. His thinking changed. His community expanded. His scope of justice and ethical concerns broadened. Proenvironmental behavior is far more likely to occur when we feel a sense of oneness with nature and a sense of a common essence that ties us together. Certainly, this is what led my colleague and me to develop the Land Ethic Scale as a measure of the land ethic worldview.

Examining this literature in more detail, Frantz and I (Mayer and Frantz 2004) reported three findings relevant to perspective-taking. Two of the findings show that the Land Ethic Scale is positively related to the ability to take the perspective of another person (r = .37 and .51). The third finding shows that the Land Ethic scale is positively related to individuals' ability to take an environmental perspective (r = .50). Subsequently, in a number of unpublished data sets, we consistently have found this scale to be associated with the ability to take the perspective of another.

Dutcher, Finley, Luloff, and Johnson (2007) explored whether greater connectedness to nature was associated with greater environmental concern. They measured environmental concern by asking participants to respond to questions like: "If things continue on their present course, we will soon experience a major catastrophe" and "The problems with the environment are not a bad as most people think" (reverse worded). As predicted, they reported that individuals scoring higher on their Connectedness with Nature Scale expressed greater environmental concern.

THE LAND ETHIC AND PROENVIRONMENTAL BEHAVIOR

Before discussing the specific details of the relationship between the land ethic worldview and proenvironmental behavior, let me present several general statements to help orient you to the details of these findings. First, as you'll see, the Land Ethic Scale and the related connectedness-to-nature scales all predict greater proenvironmental behavior. This is a robust finding. The barriers to helping covered in the last chapter seem to fall before this worldview. Perhaps this is because this worldview affects not just one stage but many stages in the Latane and Darley model. This finding demonstrates that the land ethic is an especially earth-friendly worldview.

Second, returning to a statement I made in chapter 1, this research illustrates the idea that before you can realistically expect popular support for economic interventions like the carbon tax, and expect politicians to feel pressured by the electorate to work on earth-friendly legislation, and anticipate that people will change their lifestyles to help the earth, individuals need to clearly see the importance of these changes. This research illustrates that proenvironmental actions flow from this particular worldview. It shows that an ecological worldview prompts people to take political action to make a positive change, that people who view the world through this lens are more involved in environmentalist organizations, and that they are more likely to engage in earth-friendly consumer practices and adopt a lifestyle that demonstrates their concern about the well-being of the earth. Box 4.6 below summarizes the particulars of this research.

Lifestyle/Consumer Practices

When illustrating how lifestyle changes and earth-friendly consumer practices flow from the land ethic ecological worldview, it is critical to acknowledge that if the transition to the alternative worldview of the land ethic is going to occur, people will need to be willing to change their lifestyles and consumer practices to help the environment. When considering lifestyle, it is important to note how people spend their time. Is it spent indoors or outdoors? Are they passing their time in front of a computer or a television, or are they outside in the natural world? Estimates are that people tend to spend 90 percent of their time indoors (Evans and McCoy 1998).

Frantz and I (Mayer and Frantz 2004, study 1) developed a lifestyle questionnaire to assess the extent to which participants had contact with the natural world. The first set of questions asked participants to reflect on what their "typical day" was like. They were asked to respond to statements like: "My work keeps me indoors most of the day." A second set of questions asked participants to describe how much time they spend in various locations (e.g., in front of a computer, in a car, outdoors) on a typical workday. A corresponding third set of questions asked participants to indicate how much time they spent in various locations on a typical day off. In contrast to individuals scoring lower on the Land Ethic Scale, participants scoring higher on the scale were found to spend more time outdoors and to engage in nature

BOX 4.6. HOW PROENVIRONMENTAL BEHAVIORS FLOW FROM AN ECOLOGICAL WORLDVIEW

Lifestyle/Consumer Practices

Willing to change individual lifestyle to help the environment

Likely to purchase organic food and buy products that are environmentally friendly

Likely to boycott products harmful to the environment

Likely to turn off lights and generally use less electricity

Willing to reuse things, more interested in recycling

Willing to make personal sacrifices for environmental protection

Frequently engages in nature activities

Support for and Involvement in Environmental Organizations

Likely to read a newsletter, magazine, or other publications by environmental groups

Willing to support an environmental organization and more likely to volunteer time

Willing to make monetary contributions to environmental and conservation groups

Likely to have signed petitions in favor of environmental protection

Likely to be a member of an environmentalist organization

Political Activism on Behalf of the Environment

Likely to have attended a public meeting or hearing about the environment

Likely to have contacted a government agency for information or to complain

Likely to have voted for or against candidates based on their position on the environment

Likely to support government intervention on behalf of the environment

activities, according to all three lifestyle measures (r = .55, .37, and .43, respectively).

Nisbet, Zelenski, and Murphy (2009) found that people who scored higher on their Nature Relatedness–Self Scale were also more likely to engage in nature activities. Of course, the items on several of the nature-related scales presented earlier directly assess individuals' increased activity in nature (refer back to the Environmental Identity Scale), people's desire to be in nature (e.g., the Nature Relatedness—Experience Scale), and the general joy and solace that people receive from being in nature. This lifestyle is very different from that of people who generally desire to be indoors, watching television, working on a computer, and being shut off from the natural world.

Willingness to change one's lifestyle is another important question to consider. Investigating this question, Hedlund-de Witt, Boer, and Boersema (2014) looked at the extent to which individuals reported that they were willing to modify their lives "individually through changing one's own behaviors and lifestyle." The researchers report a strong correlation between their connectedness-to-nature measure and this willingness-to-change measure, indicating people's willingness to change their lifestyle in order to help the natural world.

Using a different measure, Dutcher, Finley, Luloff, and Johnson (2007) asked participants to answer a series of questions concerning whether they were willing to change their consumer practices. The questionnaire included questions such as: Had they stopped buying a product because it caused environmental problems? Participants scoring higher on the Connectedness with Nature Scale reported engaging in these behaviors more often than individuals scoring lower on this scale.

Taking yet a different tack, Nisbet, Zelenski, and Murphy (2009) employed a measure of verbal commitment, actual commitment, and affect in relation to environmental issues concerning consumer purchases. They found their Nature Relatedness—Self Scale highly predictive of not only participants' verbal commitment to purchase more environmentally friendly products but also actual behavior related to purchasing these products, such as their willingness to purchase organic food. Moreover, participants reported a positive feeling toward engaging in these actions, indicating that this change in lifestyle is positive not only for the natural world but also for the individuals engaging in the action.

Replicating this general relationship between a land-ethic-related measure and proenvironmental consumer practices yet again, Perkins (2010) found the Love and Care for Nature Scale to be associated with a variety of these practices, such as individuals' efforts to buy products that are environmentally friendly (r = .41), to buy paper and plastic products that are made from recycled materials (r = .37), and to boycott or avoid buying products from a company that the person felt was harming the environment (r = .51).

Providing additional evidence for the land ethic worldview's association with people's willingness to change their lifestyle and engage in more environmentally positive consumer practices, Frantz and I (Mayer and Frantz 2004) initially asked participants in two studies to fill out the Land Ethic Scale, and then asked them to report on twenty-four different behaviors related to environmental protection, such as turning off the lights when a room is vacant, and avoiding the use of Styrofoam or other disposable containers. In both studies, the Land Ethic Scale was significantly related to self-reported proenvironmental action (r = .44 and .45). Moreover, in a three-year study examining actual behavior, Frantz, Mayer, Petersen, and Shamin (2013) found that the Land Ethic Scale is a good predictor of, in this instance, not self-reported, but actual, electricity use. Thus, like the Nisbet, Zelenski, and Murphy (2009) study, the Land Ethic Scale is related not just to what people say about their proenvironmental behavior (i.e., their verbal commitments) but also to what they actually do (i.e., their actual commitments). Additionally, these findings highlight the association between the land ethic worldview and people mindful of both their shopping practices and their consumption of electricity in their homes.

Perkins (2010) also found that people are more willing to change their lifestyles by making personal sacrifices on behalf of the natural world. Relating her Love and Care for Nature Scale and the Land Ethic Scale to personal sacrifices for environmental protection and conservation, participants were asked whether they would be "willing to accept cuts in my standard of living in order to protect the environment." Both the Land Ethic Scale and the Love and Care for Nature Scale were strongly associated with positive responses to this question (r = .46 and .58, respectively). Participants were also asked whether they would be "willing to pay much higher prices for many goods and services in order to protect the environment." Once again, both the Land Ethic Scale and the Love and Care for Nature Scale were strongly associated with positive responses to this question (r = .51 and .60, respectively).

Examining the Land Ethic Scale and the other related scales in her study 2, which involved a sample of participants from the United States, Tam (2013) found that each of these measures is associated with self-reported ecological behavior. The self-reported ecological behavior scale asked participants to report how often they engaged in twelve eco-friendly behaviors, such as "looking for ways to reuse things" and "recycling things (e.g., papers, cans, or bottles)." As can be seen in table 4.4, each of these measures is strongly related to self-reported behavior associated with caring for the natural world.

Support for and Involvement in Environmental Organizations

Illustrating that being more supportive of and involved in environmental organizations flows from the land ethic ecological worldview, Perkins (2010) found that the Love and Care for Nature Scale is related to interest in environmental groups, such as being open to reading a newsletter, magazines, or other publications written by an environmental group ($r = .40$) and to signing petitions in support of promoting environmental protection ($r = .44$). It is also related to currently belonging to or donating to an environmental organization ($r = .32$).

Dutcher, Finley, Luloff, and Johnson (2007), too, asked participants whether they had ever contributed money or time to an environmental or wildlife conservation group. Participants scoring higher in connectedness with nature reported engaging in these behaviors more often than individuals scoring lower on the researchers' connectedness scale.

Similarly, Nisbet, Zelenski, and Murphy (2009) found that participants scoring higher on their Nature Relatedness—Self Scale were more likely to make monetary donations to environmental groups, be members of an environmental organization, and define themselves as environmentalists.

Lastly, Tam (2013) examined the question of how the Land Ethic Scale and the other related measures were associated with attitudinal support for environmental causes. In study 2, she measured attitudinal support for environmental causes by having participants fill out a measure composed of ten questions from the Environmental Movement Activism Subscale (Milfont and Duckitt 2010). It included items like: "I would like to support an environmental organization." Examining table 4.4, you can see consistently

TABLE 4.4

Correlations between the Land Ethic Scale and Related Scales with Proenvironmental Behaviors

	Land Ethic	Emotional Affinity	Commitment to the Environment	Environmental Identity	Nature Relatedness
Attitudinal support for environmental causes	.68	.61	.73	.70	.72
Self-reported ecological behavior	.62	.54	.62	.66	.60

SOURCE: Tam 2013.

strong positive relationships between this Environmental Movement Activism Subscale and the Land Ethic Scale and other related connectedness-to-nature measures.

Political Activism on Behalf of the Environment

In illustrating that proenvironmental political actions flow from the land ethic ecological worldview, Dutcher, Finley, Luloff, and Johnson (2007) asked participants to answer a series of questions concerning whether they had ever attended a public hearing or meeting about the environment, contacted a government agency to get information or complain about an environmental problem, or voted for or against a political candidate because of, in part, his or her position on the environment. Participants scoring higher in connectedness with nature reported engaging in these behaviors more often than individuals scoring lower on their connectedness scale.

Nisbet, Zelenski, and Murphy (2009) also found that their Nature Relatedness—Self Scale predicts political activism. Similarly, Perkins (2010) found that the Love and Care for Nature Scale predicts political actions, such as willingness to vote for a candidate in an election at least in part because that candidate is in favor of strong environmental protection/conservation ($r = .48$). And Hedlund-de Witt, Boer, and Boersema (2014) found that their Connectedness to Nature Scale predicts individuals' willingness to change "societally through support for government intervention." Lastly, Tam (2013) found that the Land Ethic Scale and the other related measures are associated with "contacting a government agency to complain about environmental problems" and to "volunteering time to help an environmental group."

In summary, this research on lifestyles, consumer practices, support for environmental organizations, and political activism on behalf of the natural world highlights the fact that the land ethic worldview is associated with proenvironmental behavior in a myriad of different ways. The replicability of these findings is remarkable. Time and again, when using different scales associated with the land ethic worldview, different outcome measures, and a diverse set of participants, the same picture emerges—namely, that from this ecological worldview flows a concern for and behavior directed at creating greater harmony between human and natural systems.

Overall, this research literature illustrates that the land ethic worldview represents a less egoistic orientation than the U.S. worldview, and a more ecological, or biospheric, orientation. People seeing the world through this cultural lens express greater perspective-taking and caring for the natural world and take more proenvironmental actions. This worldview is *not* associated with a distancing of self from nature. It is *not* associated with cultural striving that harms the natural world. Rather than self-centered, it is a nature-oriented, self-transcendent worldview in which people express concern for and are not indifferent toward nature.

THE LAND ETHIC AND WELL-BEING

This less self-centered, less egoistic, and less self-enhancing orientation that characterizes the land ethic worldview is also associated with various forms of well-being. Moving beyond a concern centered on one's self-image, financial success, and status, and re-centering one's worldview on connections one has to the broader world, is a healthier orientation. After all, the mastery-oriented individualistic worldview that embraces the ideal of the autonomous self, this de-contextualized being, is a view of self that runs counter to and undermines people's basic social motive to feel a sense of belonging and connection. Moreover, via this largely mechanistic worldview, one loses both the sense of being part of a community that transcends self, and much of the love, wonder, and awe that a person can experience in nature.

Research I've been directly involved in, with both student and community samples, yields nine data sets that demonstrate significant associations between the Land Ethic Scale and individuals' degree of happiness and life satisfaction (Mayer and Frantz 2004; Frantz and Mayer 2005a, 2006; Mayer, Frantz, Bruehlman-Senecal, and Dolliver 2009). It also yields four data sets indicating that individuals who score higher on the Land Ethic Scale exhibit a greater sense of leading a meaningful life.

In study 1 with undergraduate participants from Hong Kong, Tam (2013) also examined the Land Ethic Scale and the other related connectedness-to-nature measures and compared them to measures of subjective well-being. The measures of subjective well-being she employed were the Satisfaction with Life Scale, the Affect Balance Scale, and the Subjective Happiness Scale. The Satisfaction with Life Scale (Diener, Emmons, Larsen, and Griffin 1985) is a straightforward measure that simply asks participants to reflect on their general life satisfaction. The Affect Balance Scale, by Diener and Biswas-Diener (2008), asks participants to fill out an emotional wellness questionnaire, in which they indicate how frequently they had experienced eight positive and eight negative feelings during the previous four weeks. A difference score between positive and negative feeling was then computed. As for the Subjective Happiness Scale (Lyubomirsky and Lepper 1999), participants respond to four items, such as: "Compared to most of my peers, I consider myself happier."

The findings are presented in table 4.5. The Land Ethic and the related connectedness scales significantly positively related to each of these well-being measures. To place these findings in a broader context: it is important to note that the magnitude of these correlations, although small, is similar to the magnitude of the associations between marriage and well-being ($r = .14$, reported by Haring-Hidore, Stock, Okun, and Witter 1985) and education and well-being ($r = .13$, reported by Witter, Okun, Stock, and Haring 1984). In this context, then, various factors can be viewed as contributing to overall life satisfaction, and this ecological worldview appears to be as important a contributor as other variables more traditionally associated with subjective well-being.

TABLE 4.5

Correlations between the Land Ethic Scale and Related Scales with Subjective Well-Being

	Land Ethic	Emotional Affinity	Commitment to the Environment	Environmental Identity	Nature Relatedness
Satisfaction with life	.20	.19	.24	.29	.27
Affect balance	.18	.18	.14	.29	.27
Subjective happiness	.22	.19	.20	.26	.28

SOURCE: Tam 2013.

Kamitsis and Francis (2013) reported that the Land Ethic Scale is significantly associated with psychological well-being as measured by the World Health Organization Quality of Life Scale. Additionally, Leong, Fischer, and McClure (2015) found that the Land Ethic Scale is positively associated with a general health questionnaire. In their study, participants were asked to rate their feelings over the previous several weeks, answering questions about whether they had, for example, "lost much sleep over worry." They also completed a questionnaire about mental health—which focused on positive aspects of mental well-being, such as "I've been feeling relaxed" and "I've been thinking clearly"—as well as a positive mood measure. That the land ethic worldview was positively correlated with these measures emphasized that this worldview is associated with a higher quality of life, less stress, a clearer focus, and happiness.

Poon, Teng, Chow, and Chen (2015) found support for the notion that connectedness to nature contributes to personal well-being by fulfilling the basic social motive for belonging. This interesting point is worth considering for a moment. Researchers have typically thought of this motive being fulfilled through positive interactions *with people*. The idea that feeling at home in nature, as part of a community with nature and emotionally tied to the natural world, can fulfill this basic need tells us that feeling connected to nature can serve as a basic wellspring for well-being.

Zelenski and Nisbet (2014), in both a student and a community sample, found that their Nature Relatedness Scale and Schultz's Inclusion of Nature in Self Scale predict participants' levels of happiness, life satisfaction, and positive affectivity. Similar to what my colleague and I discovered, they also found that the Nature Relatedness and Inclusion of Nature in Self measures are associated with community participants' experiences of a greater sense of *purpose* in their lives. Considering other positive aspects, Zelenski and Nisbet (2014) reported that the connectedness measures they used are positively related to personal growth and vitality.

Accentuating the relationship between connectedness and positive human functioning, Howell, Dopko, Passmore, and Buro (2011) related the Land Ethic and Nature Relatedness Scales to *psychological, social, and emotional well-being*. In their study, *psychological well-being* refers to ratings of *personal growth, purpose in life, self-acceptance, and positive relationships with others*, while *social well-being* refers to ratings of *social acceptance, social actualization, social contribution, social coherence, and social integration*.

Examining their findings, they reported that the strongest relationship that exists between the Land Ethic Scale and the Nature Relatedness Scale results from the psychological and social well-being measures. It's useful to consider several important aspects of these findings, beginning with psychological well-being. The relationship of these measures with personal growth, experiencing a greater purpose in life, and greater self-acceptance corroborates and extends previous findings. Second, the relationship of these measures with individuals reporting that they have more positive relationships with *other people* provides another interesting extension of this research. Life on Eaarth will require people to cooperate with one another. They will need to work together to face a wide variety of challenges. More positive relationships with others may be essential to make this happen.

The relationship between these scales and the social well-being measure is also fascinating. It highlights the fact that the land ethic worldview is associated with individuals who experience greater social acceptance and trust in others. Such people are more likely to have a positive view of human nature and feel comfortable with others. They are more likely to see others as capable of kindness and as industrious. They tend to be more hopeful about the future of society (social actualization). They see their personal lives as understandable and as possessing purpose and meaning (social coherence). They have a personal sense of sharing something in common with others and see themselves as meaningful members of society (social integration). They also see themselves as vital members of society who can make a positive difference in the world (social contribution).

In other words, the land ethic worldview connects people to nature *and to social others* in a positive manner. It promotes a stronger sense of community between self and nature and between self and other people. It enhances individuals' sense of self-efficacy by encouraging them to feel they can make significant social contributions. Stronger communities and maintaining a sense of self-efficacy will play vital roles as we face the climate change challenges ahead of us.

Capaldi, Dopko, and Zelenski (2014), in a meta-analysis based on thirty samples (n = 8523) of the relationship between nature connectedness and happiness, reported a small but significant effect size (r = .19). They noted that vitality had the strongest relationship to the connectedness measures (r = .24), positive affect the second strongest (r = .22), and life satisfaction the third strongest (r = .17). As they state, "The results suggest that closer human-nature relationships do not have to come at the expense of happiness. Rather, this meta-analysis shows that being connected to nature and feeling happy are, in fact, connected" (p. 1).

In summary, the land ethic worldview is associated with many positive aspects of human functioning. People who see the world through this cultural lens tend to not feel isolated, but instead see themselves as vital members integrated into the social fabric of a society. They tend to see their world as predictable and their lives as meaningful. They see themselves as able to make positive changes. Moreover, they exhibit vitality, self-acceptance, a general positive emotional orientation, and positive relationships with others.

Regarding the question of whether to switch to this worldview, one main issue is the cost involved in giving up one worldview in order to transition to another. From my perspective, the cost seems minimal. As described in chapter 2, there are clear distortions and shortcomings associated with the mastery-oriented individualistic worldview, including feelings of isolation and the false promise that materialistic wealth leads to happiness; individuals who embrace this worldview may find themselves on a treadmill to nowhere. Transitioning to the land ethic worldview would not only promote greater harmony between human and environmental systems but also enhance individuals' sense of well-being. In fact, the real "cost" that I see lies in *not* switching to the land ethic worldview and in blindly following the mastery path to a potentially tragic end. As Nisbet and Zelenski (2011) state, there may well be a "happy path to sustainability."

THE LAND ETHIC AND PROBLEM SOLVING, HOLISTIC THINKING, AND CREATIVITY

Besides being associated with proenvironmental behavior and well-being, the land ethic worldview has been associated with more effective problem-solving behavior. For instance, my colleagues and I (Mayer, Frantz, Bruehlman-Senecal, and Dolliver 2009) asked participants to think of a problem they faced in their lives and then try solving that problem. In an experimental setting in which we manipulated the land ethic worldview by placing half of the participants in a natural environment and half in an urban environment, we found that the land ethic worldview was associated with both heightened well-being and an enhanced ability of participants to solve the problem they focused on.

Why might the land ethic worldview be associated with enhanced problem-solving ability? Leong, Fischer, and McClure (2015) found that the Land Ethic Scale is associated with holistic thinking and creativity. Their findings show that people seeing the world through this lens are more likely to see connections and interrelationships. Higher Land Ethic Scale scores were associated with people being more innovative and full of ideas.

Nisbet, Zelenski, and Murphy (2009) found that their Nature Relatedness—Self Scale is associated with greater consideration of future consequences, which refers to "how people differ in their contemplation of how potential current behavior may affect future events (Strathman, Gleicher, Boninger, & Edwards, 1994)." Their findings emphasize that holistic thinking not only enhances the ability to see more connections in our present circumstances but also creates the ripple effects of our actions through time. To cultivate sustainability, people need to be more creative. They need to be aware of the short-term and long-term consequences of their actions. That this ecological worldview leads to this type of thinking is promising. Thinking ahead might enable people to avoid future problems.

THE LAND ETHIC AND SPIRITUALITY

As for why the land ethic is associated with well-being, it may well be that multiple factors are contributing in unison to this relationship. A greater sense of personal self-efficacy, a greater problem-solving ability, and fulfillment of the need to belong may all play a role. Another factor, however, is spirituality. Recall from earlier in the chapter that a spirituality question was part of several of the connectedness-to-nature measures. That the egalitarian and harmony value domains comprise the self-transcendent dimension of the Schwartz model also speaks to the possibility of spirituality playing an important role. In other words, the land ethic worldview appears to be associated with more than simply a sense of kinship and communality with the natural world. The research argues that this sense of oneness with the natural world can also be experienced as a spiritual connection to nature characterized by a sense of wonder, awe, and sacredness. This spiritual sense of connectedness to nature involves what some call being connected to the "community of creation" (Woodley 2012). The relationship between the land ethic and spirituality is underscored by the research of Hedlund-de Witt, Boer, and Boersema (2014) and Kamitsis and Francis (2013).

Hedlund-de Witt, Boer, and Boersema (2014), using their Connectedness to Nature Scale, investigated whether a sense of oneness with the natural world was associated with a measure of spirituality, using items like: "I have sometimes had experiences that you could call spiritual," "I see the earth and humanity as part of an ensouled or spiritual reality," "I believe every human being has a spiritual or divine core," and "I believe the universe gives expression to a creative intelligence." They found that connectedness to nature was more strongly associated with this contemporary spirituality measure ($r = .41$) than it was with a measure of a traditional God ($r = .28$).

Even more interesting is a study by Kamitsis and Francis (2013), who employed the Land Ethic Scale and a measure of spirituality composed of items like "I have had an experience which I knew to be sacred" and "I have had an experience in which something greater than myself seemed to absorb me." They investigated how these two scales were related to participants'

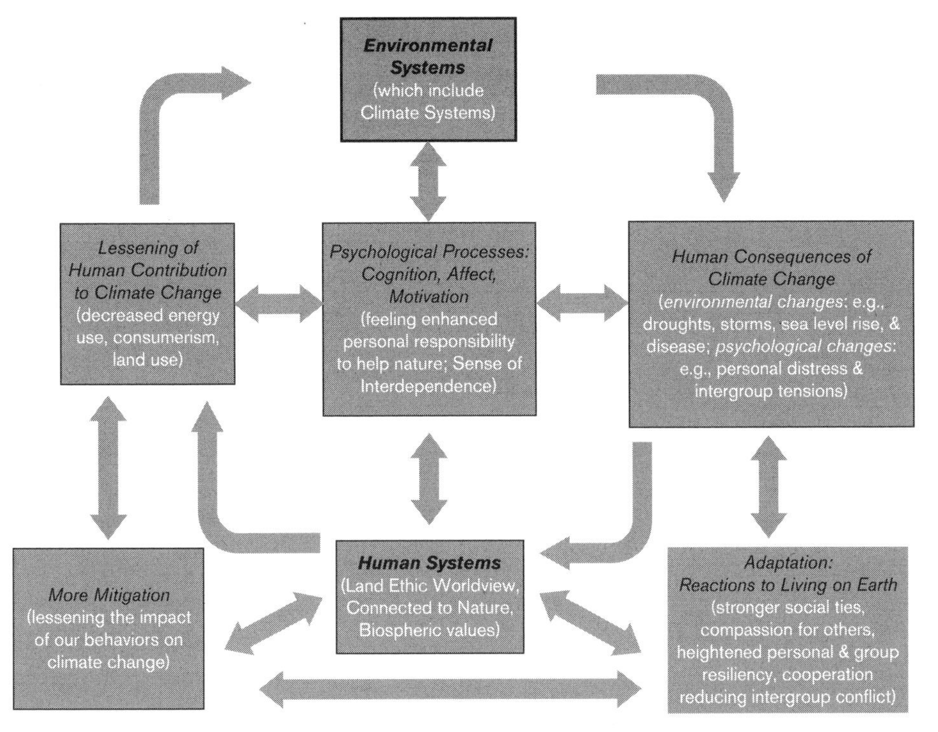

Fig 4.4. Adapted from Swim et al., "Psychology's Contributions to Understanding and Addressing Global Climate Change," *American Psychologist* 4 (2011): 242, fig. 1.

scores on a well-being scale. What they found was that spirituality mediated the relationship between the Land Ethic Scale and well-being. In other words, they found that the Land Ethic Scale was associated with well-being only to the extent that it also influenced spirituality. This is only one study, so it is difficult to dismiss all the other reasons for why the land ethic might be associated with well-being; but nonetheless this is a provocative finding.

FINAL THOUGHTS

Creating Greater Harmony between Human and Environmental Systems

When we examine the impact of human systems from a land ethic worldview, things look decidedly different than when regarded from the mastery-oriented individualistic worldview, which we viewed earlier. Here we see that this ecological worldview leads to an enhanced sense of personal responsibility to care for nature. We see a lessening of human contributions to climate change. We see more mitigation. As for adaptation, we find that stronger social ties, greater compassion for others, and cooperation create a picture starkly opposed to images associated with the mastery-oriented individualistic worldview.

That the land ethic worldview connects people to the natural world and to other individuals speaks volumes about how this worldview orients people to the world in a more caring and positive manner than the mastery-oriented individualistic worldview does. That the former is associated with a sense of personal vitality, self-efficacy, personal growth, and a greater sense of awe and wonder speaks not just to the self-transcendent nature of this worldview but to the self-transformative nature of it as well. Seeing the world through this cultural lens certainly is about creating greater harmony between human and environmental systems; but as becomes evident as this story unfolds, it is also about how we can be better, healthier, and happier individuals.

Creating Greater Harmony between People

The "adaptation" part of the model in figure 4.4 concerns our reactions to living on Eaarth. That the land ethic worldview is related to greater perspective-taking, not just from the perspective of the natural world, but also from the perspective of other people, suggests that this worldview should be associated with greater helping and caring for others. I think the Howell, Dopko, Passmore, and Buro (2011) study is particularly interesting in this regard. They found that the Land Ethic and Nature Relatedness Scales relate to individuals' sense of having positive relationships with others, to seeing themselves as meaningful members of society (social integration), and to seeing themselves as vital members of society who can make a positive difference in the world (social contribution). These findings speak to the possibility that this worldview is related to greater harmony within human systems. That there is greater spirituality associated with this worldview and less egoistic striving also points to the possibility that this worldview enables people to sense a special connection with nature and other people, which fosters a general caring orientation and greater resilience. In life on a harsher world, these positive contributions of the land ethic to interpersonal and group relations may play an important role.

Five decades of research in social psychology has demonstrated how feeling a sense of interdependence with others promotes intergroup relations (cf. Pettigrew and Tropp 2011). Being able to take the perspective of others, having positive relationships with others, and working with others have all been shown to break down barriers between people and promote intergroup harmony. That the land ethic worldview has been shown to reduce the human/nature split and appears to be related to reducing barriers between individuals and groups has critical implications. It suggests that this worldview is associated with promoting a broad-based harmony not just between humans and nature but within human groups as well.

The Value of Interconnectedness

In fact, one interesting model in peace psychology places interconnectedness at the very center of things. As you can see in the Teixeira (1999) model in figure 4.5, interconnection is at its heart. In other words, from this perspective, in order for us to create a more peaceful world, our thoughts, beliefs, feelings, and actions all need to derive from a more holistic perspective, where we see the interconnections around us. Seeing the world through the lens of interconnection will then help us take the perspective of others, feel compassion, and experience greater harmony within ourselves (intrapersonal). Additionally, this lens leads to greater harmony between individuals (interpersonal) and, even more broadly, beyond human encounters (transpersonal), which includes our experiences with nature. As illustrated in this model, if we view the world through the lens of interconnection, the broader arena of how we go about childrearing, education, conflict resolution, social struggle, and confrontation will be affected. The sense of interconnection radiates to and affects the very nature of what we view favorably and reward. In this way, in time, this sense of connection will transform our way of life.

Returning to Diamond's (2005) psychological factors related to societal collapse that I presented in chapter 1, recall that he states that societies may collapse because the group may fail to "anticipate a problem before the problem actually arrives." He also points out that a society may collapse because

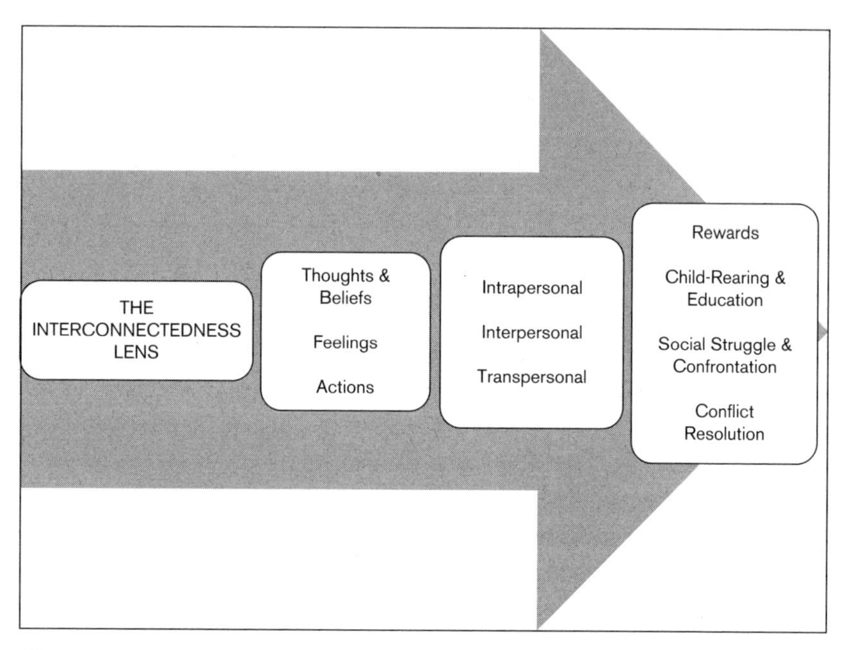

Fig 4.5

the problem arrives and they fail to see it. Moreover, he emphasizes the importance of intergroup relations and how a society may collapse owing to hostile neighbors or the lack of friendly trade partners. Regarding how switching to a land ethic worldview addresses these concerns, the research I've presented in this chapter points out that a sense of interconnectedness leads to anticipation of future events, enhanced problem-solving ability, and creative solutions. Another potential benefit of transitioning to this alternative worldview, as highlighted by the findings of this research, is that a sense of interrelatedness might lead to friendlier, or at least less hostile, intergroup relations between nations and international trade partners.

Reflections on the Great Transition

The Great Transition conveys a sense of a moving from here to some other place. The reality of it, however, is that we probably all have some aspect of this worldview already present within us. This transition may well be a matter of accentuating, bringing to the foreground, and nurturing a vision that we all hold to a greater or lesser extent and, simultaneously, diminishing or casting off a cultural lens that no longer is useful. The Great Transition probably doesn't involve going anywhere other than into our thoughts and feelings, making accessible in our minds a sense of connectedness that may lie buried. It may be nothing more than saying to oneself, "I'm tired of this lifestyle," and desiring to nurture a different way to see, and live in, the world.

Of course, this is probably easier said than done. However, my experience and my research have taught me that it is literally done by thoughtfully choosing to take one step after another in a new direction (Mayer, Duval, and Duval 1980). It involves waking up in the morning, looking in the mirror, and saying to oneself that today or even this morning, "I will take steps that are different from the steps I've taken on previous days and mornings." It involves a mindful routine of doing this deliberate act each morning, each week, and each month until these new steps become habitual and replace the old routines of the past. It might involve a daily walk in the woods. It might involve a walk on the beach or spending time with a pet. Although old habits die hard, new habits can be created; and one important lesson from psychology is that it is not just our thoughts and feelings that lead to new behavior. Our behavior, too, can lead to new thoughts, feelings, and a new way of seeing the world.

A Word of Caution

The question of whether we can transition to the land ethic worldview is only part of the issue. Another critical question is whether we can make this transition in a timely manner, for the clock is ticking and we need to avoid letting climate change reach a tipping point.

A related issue concerns how our present life on Eaarth may make it more difficult for people to find solace and comfort in nature. While living in the more benevolent climate of the Holocene period, people found that feeling renewed and positive in nature was relatively easy. Living on Eaarth during the Anthropocene period, however, we may find it more difficult to feel connected to a natural world that is harsher and less predictable. People may also suffer from the loss of members in their expanded community. In other words, the transition to a land ethic worldview needs to occur sooner rather than later.

Can It Be Done?

So, when looking at the night sky, is it the darkness or the stars that you see? I think of the research described in this chapter as the points of light that can guide us toward a brighter future. Although change is hard and there is an imminent climate threat before us, the good news is that cultural change is already occurring. Our cultural course can change and is changing. We can collectively plot a wise course for our greater cultural ship and avoid a *Titanic*-like tragedy. Chapter 5 illustrates the many ways in which this transition is taking place. The purpose of this final chapter is to provide you with the hope that we can make this transition. The examples are also meant as a "how to." In order to successfully make this transition, individuals and communities need not figure it out for themselves. There are examples and organizations already present that provide guidelines for engaging in this transition.

CHAPTER FIVE

Actions Being Taken to Transition to the Land Ethic Worldview

Faced with a problem, we must see it clearly before taking action. Without a clear vision of why a problem exists, we cannot formulate a clear set of goals and course of action. This sentiment is captured in Kurt Lewin's famous saying "There is nothing as practical as a good theory." For Lewin, a theory is like a map. Without a good map, how can you hope to arrive at your destination? Up to this point, I have laid out the map and pointed to a destination. I've articulated the contours of the mastery-oriented individualistic worldview, the distortions associated with it, and its negative influences on our relationship to the natural world, and I have pointed to an alternative destination. Namely, I've argued that we need to adopt a land ethic worldview, which will enable us to see connections instead of divisions, experience empathy for others and the natural world instead of indifference, and act in beneficial instead of harmful way toward both nature and other people. This is the clearer vision. With map in hand, I illustrate in this final chapter the different path. People and organizations are using this map to actively engage individuals and communities in transitioning to a new way of seeing and relating to the world.

In a race against the clock to avoid the worst that climate change has to offer, people and communities are stepping up, getting involved, and making a difference. In the absence of effective federal leadership, a worldwide grassroots movement has emerged. Against the tide of hyperindividualism, people are making connections to one another and to the natural world.

Make no mistake: the transition to a land ethic worldview is taking place. There is no need to say that the task is impossible, for it is occurring.

While not every project I present here explicitly uses the term *land ethic* to describe its goals, implicitly these projects aim to connect people to one another and to nature. This is what the land ethic is about. As we saw in chapter 4, the land ethic worldview and the feeling of connection to nature are almost synonymous. Moreover, connectedness to nature is associated with a host of positive outcomes that extend well beyond proenvironmental behavior. Hopefully, by reading this documentation of the movement now taking place, you'll not only become more optimistic that the necessary changes can take place but also will find ideas for how you and others might contribute to these changes.

THE OBERLIN PROJECT

The Oberlin Project is a prime example of a community transitioning to a future where the psychological distance between individuals and nature has diminished, and where the community treads lightly on the natural world. One of the main aims of the project is to demonstrate to other towns and cities that this transition can occur. Regarding the land ethic worldview, it is clearly present in the way the project has been conceptualized, designed, and implemented. The project underscores the fact that to confront the environmental threat of climate change, we need to shift from the mastery-oriented, individualistic, analytic, and piecemeal style of thinking that gave rise to our present predicament, to a more egalitarian, earth-centered approach. We must also recognize our interconnectedness with the natural world and think in more holistic ways in order to confront the threat.

Oberlin is a small town of approximately eighty-two hundred people, located thirty-five miles southwest of Cleveland, Ohio. Oberlin College, founded in 1833, is situated in the center of the town. The college, one of the most respected of the elite liberal arts colleges in the nation, is known for being the first college to admit and graduate African Americans and women. Before the Civil War, the town was known as the northernmost stop on the Underground Railway. Both the college and the town have long been in the forefront on progressive issues. Today, this proud heritage continues as the community sets an example of how to transition to a new way of seeing and being in relationship with the natural world.

David Orr is the visionary behind the Oberlin Project. Hired by the college in 1990 to head the environmental studies program, he is an influential educator, writer, and speaker on issues related to environmental sustainability. He also raised money to build the Adam J. Lewis Environmental Studies Center. A historically important building, the first substantially green building on a U.S. college campus, it was designated by the U.S. Department of Energy as one of the thirty milestone buildings of the twentieth century. Powered entirely by sunlight, designed and built in an earth-sensitive manner, and incorporating biophilic design features, such as windows and a waterfall in the lobby that tie people to the natural world around them, this building demonstrates that the built environment doesn't have to be wasteful or disconnect people from nature. In many ways the Oberlin Project has taken the lessons learned from designing this building and increased the scale of the overall endeavor to include the entire community.

Illustrating that the project emphasizes holistic thinking, both the city and the college are involved in this model of "full spectrum sustainability. . . . In short, we aim to join issues normally kept separate into a system in which each of the parts reinforces the health and resilience of the larger community" (Orr 2014, p. 10). Issues surrounding the local food economy, energy, policy and finance, community economic development, and education have all been linked with one another. The project is designed "to give practical meaning to the idea of systems in the day-to-day affairs of the City, the College, and the local economy" (Orr 2014, p. 10).

Working groups concerned with education about sustainability, local food systems development, local land and agriculture practices, community engagement, transportation, economic development, and creating a climate-positive community all work separately and together in a holistic manner to make their vision of the project a reality. The project's organizers recognize that if you want a climate-positive, postcarbon community, where no one relies on fossil fuels, you need to educate people to achieve it. Moreover, education must be linked with jobs. A new energy economy, one not based on fossil fuels, has to be established. Additionally, the community has realized that you can't keep shipping in food from an average of fifteen hundred miles away, since that needlessly contributes to CO_2 emissions and makes a community overly reliant on distant places for their food. Instead, local farms must be supported and a new generation of farmers trained. Perhaps most importantly, community engagement needs to be nurtured, since the goal is for the community to take ownership of the project and, in an egalitarian manner, continue to grow it. All of these elements have been linked with one another, and Oberlin has made substantial progress. For instance, as a result of moving away from reliance on fossil fuels and toward a postcarbon future, 87 percent of Oberlin's electricity comes from renewable sources at present.

Linking education about sustainability, local food systems development, local land and agricultural practices, and economic development with one another, these working groups aim to create a community where local farms supply fresh produce to Oberlin. The project supports these local farms in the form of training. One training facility is the George Jones Farm, located at the edge of town. As stated on their website, "At the Jones Farm we 'grow' farmers through Intern-Apprenticeships with Oberlin College students and provide a living laboratory for students in the Lorain Community College Certificate of Sustainable Agriculture Program. Food is grown for City Fresh [which provides fresh, organic food to the "food deserts" in the Cleveland area] and Oberlin College. Programs for schoolchildren and three weeks of Summer Discovery Camp provide children with the opportunity to learn how food grows and experience a variety of outdoor activities."

To elaborate on the Lorain Community College Certificate of Sustainable Agriculture Program: this local community college provides another educational program that "grows" farmers. As their website describes it: "Sustainable agriculture is one that follows the principles of nature to develop systems for raising crops that are, like nature, self-sustaining. Sustainable agriculture's success is indistinguishable from vibrant rural communities and rich lives for families on profitable farms that provide for current and future generations." This is a two-year certification program that enables students to then gain support from an entrepreneurial program to put their education into practice. In other words, the Oberlin Project does not simply educate students about sustainability and local land and agricultural practices. It's also about economic development. The vision for the Oberlin Project is to surround the town with small, productive, economically viable farms that produce much of the town's food.

Besides selling their produce to City Fresh and Oberlin College, students at George Jones Farm take produce each Saturday to Oberlin's farmer's market, another outlet where local farmers and artisans sell their goods. Shopping at this market is an experience very different from shopping at the Walmart south of town or traveling to the Heinen's grocery store in a neighboring community. The market provides high-quality food and craft items to the community, but there's more to it than that: seeing people mill around the market on a Saturday, talk with one another, and share stories highlights the community-building aspect of the market. In contrast to relying on their cars to shop at markets outside of town, people often walk or ride their bikes to this market. During much of the year, residents and others can attend this outdoor market, experiencing both the seasons and the foods linked with specific seasons, which ties people to the natural rhythms of the earth. Eating the fresh and flavorful food at home, in contrast to the often flavorless food one can buy at a chain grocery store, provides another link between people and nature.

Like the college's programs designed to "grow farmers," school programs encourage students at all levels, from kindergarten through college, to recognize their interconnectedness with the natural world. In the early grades, students plant class gardens. This activity extends the classroom to the outdoors and engages the students in watering and caring for their plants. Schoolchildren take field trips to the George Jones Farm, which is also a nature preserve. On this seventy-acre piece of land, children learn how food grows. They actively explore the nature preserve, too, which includes wetlands, forests, and an abundance of wildlife and native plants.

In each school in Oberlin, "environmental dashboards" are prominently displayed. The dashboard is a large television monitor that portrays moment-by-moment energy use in a building, which illustrates the relationship between energy use and CO_2 emissions. This feature of the dashboard is intended to educate and empower students by directly showing them how their actions affect the world around them. Moreover, they can see how these emissions are tied to the threat of climate change. Thus, the dashboard's real-time feedback enables students to not only see how their actions affect emissions but also experience, through trial and error, how they can effectively take action to reduce their emissions.

The environmental dashboard has another feature designed to increase individuals' sense of connection to nature and to help transition them to a land ethic worldview. This feature illustrates a person's embeddedness in the natural world. It presents an abstract image of a person located in Oberlin, within the context of identifiable aspects of the natural world that surrounds her. It depicts Plum Creek, which runs through town, the Black River, which the creek drains into, Lake Erie, and the local woods and fields. In other words, it places the individual within the local bioregion.

Dashboards that include one or both of these features are in our local schools and a number of other public and private spaces around town, such as our public library, a favorite coffee shop, and many condos and homes. Working with John Petersen and Rumi Shamin of the environmental studies department at Oberlin College, Cindy Frantz and I created these images to increase individuals' experience of feeling connected to nature and to promote their holistic thinking. Our research demonstrates that these images placing a person in the local bioregion can contribute to individuals' transition to a land ethic worldview.

Education about sustainability is linked to economic development in another way, too: at Oberlin College, the local community college, and the joint vocational school just south of town, students learn skills that enable them to create smarter homes—that is, homes that use less energy. This training involves teaching students how to effectively insulate homes, minimize electricity use in a home, install environmental dashboards, and efficiently use water inside and outside the house. Students are also trained to install programmable thermostats. As this technology develops, students will learn how to teach people to use smartphones and other electronic devices to control these thermostats from afar.

Students are trained to work on transportation issues, learning, for example, how to enable citizens to be less car-dependent, how to repurpose an engine to run on biofuel, and how to increase bicycle use in the community. Ultimately, the goal of the educational experience is to create new jobs for students in our community. Ideally, in the years to come an ever-increasing number of jobs and businesses will take root in Oberlin. Our young people will find that they don't have to leave the community in search of work. They can stay and raise their families in this town. As their place attachment grows, their caring for others in the community and for the natural beauty of the town can grow. Thus, the Oberlin Project is not just about creating an earth-friendly town but also about building a stronger community.

A transportation option already available was developed through a partnership between Oberlin College and Enterprise car rental. Named "WeCar by Enterprise," this transportation option enables members of the community to rent a fuel-efficient car by the hour, for the day, or overnight. This is one program designed to enable citizens to give up their personal car ownership, which will reduce CO_2 emissions and have the added benefit of reducing the need for parking spaces and parking lots in town. Less asphalt for parking translates into more green space. The town has also created more bike lanes and convenient parking areas for bikes. That a beautiful, converted rail-to-trail bike path runs through the middle of the town adds to the allure of owning a bike and using it.

The egalitarian aspect of the land ethic is further fostered by community involvement in the implementation of the Oberlin Project. From the very beginning, focus groups were convened throughout the community, and input from community members was encouraged. Community involvement has continued in the form of electing city officials with different ideas about how to implement the project. Discussions between council members and community members about the project often occur at council meetings. Open forums at the college and letters to the editor in the local paper are other ways that community members contribute ideas.

Oberlin community members are further encouraged to be involved in the project by participating in city utility programs such as free energy audits. These audits are designed to enable all homeowners in town to have their homes assessed for energy efficiency, and utility employees recommend ways to increase energy savings. Assistance is also provided to low-income families to increase the energy efficiency of their homes at no personal cost. "Going green" is thus not just a privilege of the wealthy. All are encouraged to participate.

Overall, the Oberlin Project exemplifies a transitioning to a land ethic worldview through its planning process and the nurturing of programs aimed at connecting people to nature and to one another. It involves educational experiences that translate into actual jobs. It involves community participation and offers opportunities for everyone to be involved. You can see in Oberlin a community transitioning to a new way of thinking about the problem of climate change and then living within this new vision.

Getting people actually out in nature has to be a critical aspect of any transition to a land ethic worldview, however. Our research strongly argues that this is the most effective and direct way to increase Land Ethic Scale scores (see Mayer, Frantz, Bruehlman-Senecal, and Dolliver 2009). Certainly having children in early grades working in outdoor gardens helps. Having them take field trips to the George Jones Farm and nature preserve is also a step in the right direction. Making use of the farm as a place for summer camps provides another meaningful outdoor experience. The feature of the environmental dashboard related to increasing connectedness to nature also has been demonstrated to be effective. But when the children go home, do they sit in front of a television or computer? And when adults leave the coffee shop or the public library, where the dashboard is present, where do they go?

With research showing that people spend the large majority of their time indoors, how do you encourage them to go outdoors? In answering this question, we need to keep in mind two main considerations: (1) how accessible is the green space, and (2) how rewarding is it to be in this space? Do people have to drive forty minutes to get to it, or is it down the block from where they live? Is the space a littered patch of brown grass, or is it an interesting place to be? In Oberlin, we are fortunate to have a thirteen-acre town square as the central point for both the college and the town. This square, planted with different types of trees and other vegetation is a beautiful green oasis, and college students, families, couples, and youth can always be seen making use of the space. Throughout the year, different community festivities take place on the square. During summer, concerts held on the square attract large audiences from inside and outside Oberlin. Overall, providing this accessible and attractive natural setting is a critical aspect of the transitioning process.

Another accessible and attractive green space in Oberlin is the arboretum. Composed of a ninety-four-acre wooded area with trails, a creek, and two ponds, "the Arb" is only three blocks from the middle of town.

Bird-watchers and runners frequent the woods. Risk-takers trying their luck by walking on a log over the creek can be found there. Dog-walkers and people who simply like to stroll in the woods make a daily habit of walking the paths. The ponds provide opportunities for kids accompanied by parents or grandparents to try their luck at fishing. Others may simply opt for sitting on a bench overlooking the water with the woods framed behind it, taking in the natural beauty of this spot.

A recreation facility with a beautiful outdoor pool and nearby baseball fields and soccer fields provides a third spot for children, youth, and adults to enjoy the outdoors. In the heat of summer, hordes of kids make use of the outdoor pool. Parents with their toddlers occupy one section of the pool, while the teenagers frequent the deeper part and the water slide. In fall the town soccer league for children in kindergarten through middle school claims the soccer fields, while in spring and summer the baseball and softball players take over the fields. And, as I mentioned previously, the town has a beautiful and well-maintained bike path that runs through the middle of town and out into the surrounding countryside. Bicyclists, runners, and walkers all make the most of this trail, as do cross-country skiers in winter.

While the transition to a land ethic worldview and postcarbon community is a critical issue, the goal of creating a resilient community must also be addressed. With an uncertain future before us, and with climate change already taking place, communities need to be strong. They need to be able to withstand the possible shocks of climate change. They need to be able to bounce back. Not depending on long supply chains for food, and focusing instead on locally grown food, should help Oberlin if megadroughts compromise food production in other parts of the United States. By developing its own energy sources in the form of solar panels, geothermal heat pumps, and methane captured from the local landfill, the town should experience less of an impact from disruptions of the national energy grid. By transitioning to a land ethic worldview, which, as we saw in chapter 4, is associated with proenvironmental and greater social cohesiveness, problem-solving skills, creativity, and well-being, the community should also be better able to absorb the climate-related blows—that is, not splinter, but remain intact. Neighbors will be more likely to help neighbors.

TRANSITION TOWN TOTNES

Transition Town Totnes and the Oberlin Project have many similarities. Connecting people to nature and thinking holistically are at the heart of this project. Moreover, this project has a clear egalitarian emphasis and a human/nature spiritual flavor to it. Before getting more into the details, however, let me provide more of a context.

Totnes is a town of approximately eighty-one hundred residents located 220 miles southwest of London in Devon. A beautiful town on the river Dart, a few miles from the coast, Totnes is known for featuring music, art, theater, and natural health. It is also known as a progressive town, one reenvisioning its future and actively transitioning away from dependence upon fossil fuels.

Rob Hopkins, an environmental activist and writer, is one of the founders of the Transition Town movement, which has become a worldwide movement (see Hopkins 2008, 2011; Chamberlin 2009). Totnes provides a good example of this movement, and while it is similar in many ways to the Oberlin Project, at the same time it does have a different emphasis. Most importantly, while the issue of resilience is more of a subtext for the Oberlin Project, it is placed front and center in Totnes. A main goal is to prepare the town for the inevitable environmental shocks associated with climate change and, with oil being a finite resource, for the inevitability of fuel shortages at some point in the future.

Developing a self-sufficient community is one of the project's major goals. As in Oberlin, the residents of Totnes emphasize providing locally grown food and developing local energy sources, such as biofuel and solar energy. Counter to the cultural trend that big is better, small is good from their perspective. Keeping the goal of resilience in mind, and not depending on larger food and energy transport systems, will buffer the community from shocks if these larger systems become disrupted.

Education about sustainability and resilience are at the heart of the project. Thinking that environmental shocks will only increase over time, schools in the project teach students about conflict resolution. Given that shocks may very well disrupt outside sources of support, people are being instructed on practical skills that keep a community functioning, such as gardening, carpentry, and skills relevant to creating biofuels and repurposing engines to run on biofuels. Counter to the general trend to eliminate shop classes and classes associated with cooking and clothing construction, which have increasingly become looked down upon, in Totnes these skills are valued. They are seen as necessary skills that members of a community need in order to maintain its healthy functioning.

Totnes is an interesting transition community because members of the community are actively involved in visioning the future of the town. In the face of environmental threats, and with the realization that the old cultural narrative of materialism, wealth, and hyperindividualism has largely caused environmental degradation and not the promised happiness, community members have been encouraged to envision a new cultural narrative, a positive vision of the future that embraces the connection between humans and nature. As Hopkins states, "It is one thing to campaign against climate change and quite another to paint a compelling and engaging vision of a post-carbon world in such a way as to enthuse others to embark on a journey towards it" (2008, p. 94).

Examples of these visions highlight the connectedness between humans and nature. Through the years, Hopkins has considered himself to be something of a "vision-harvester." For instance, Stephan Harding, who lives in the Totnes area and wrote *Animate Earth*, shared a vision with Hopkins of a world where people are reconnected with nature, where there would be "an interconnected network of ecovillages, with lots of wild countryside in between, but also some lovely small cities where there would be theatre, culture, museums and good libraries, and good coffee shops, gorgeous organic architecture" (Hopkins 2008, p. 102). Illustrating the egalitarian nature of one vision, Hopkins recounts a conversation he had with Brian Goodwin, author of *Nature's Due*. Hopkins reports Goodwin's idea of a world where "humanity becomes, as he [Goodwin] puts it, 'largely invisible'; that is, more blended into and in tune with our natural surroundings. He told me, 'I'm not talking about a Rousseau 'back to nature'; I'm talking about using appropriate technology, natural materials and energy to achieve a lifestyle in which we blend with the natural world. We will have learned how to live in a way that other species have, and therefore have reduced our footprint, decreasing it dramatically to the point where we are one among many instead of an absolutely dominant species" (Hopkins 2008, p. 102). These are rich visions of potential futures that capture our imagination. That community members can hear these visions, create their own ideas, and share their dreams with one another highlights their positive approach and the egalitarian spirit that infuses the project.

It helps that an environmental think tank, Schumacher College, is located on the western edge of the town, on the Dartington Hall Estate. This college is named after the famed British economist E. F. Schumacher, who wrote the highly influential book *Small Is Beautiful: A Study of Economics as If People Mattered*. At this picturesque college, which specializes in environmental education, students can earn a master of science degree in holistic science, a master's degree in economics for transition, and a master of science degree in sustainable horticulture and food production. According to Bill McKibben's characterization of the place, "Schumacher College is a very special place. As we try to figure out what on earth we're going to do with this unraveling planet, it's become a thinktank of hope, a battery for positive vision!"

In various ways, the college teaches students how to reconnect with nature and how to envision humans living in harmony with nature. As Tim Crabtree, a senior lecturer at the college, wrote in a (now defunct) blog on the Schumacher College website: "At its heart, the College is about transformative education . . . and there is a central focus on both outer and inner transformation." New ways of living on earth are explored. In many instances, students delve into rediscovering how ancient practices can inform our present way of life. The emphasis of the education is that sustainability is more than simply a technological problem.

The transformative aspect of the educational experience is very much related to changing the way people see the world. The inner transformation involves self-transcendence. Elaborating on this, Crabtree goes on to say that the educational experience taps into "the eternal human yearning to be connected to something larger than one's own egos," which is framed as a spiritual issue. Thus, as Crabtree views education at the college, it is not only about seeing "beyond the notion of sustainability as a technological problem," but also about seeing "sustainability as a spiritual challenge."

Schumacher College is also very much about building community and creating an egalitarian way of life where each member of the community feels respected and cared for. Students and staff work in teams to take care of the day-to-day responsibilities at the college. One day a team might be responsible for cooking the meals, while another day they might tend the garden or clean the facilities. Community meetings and group festivities are another way that members of this community are brought together and individual ties are strengthened.

The teaching at the college very much informs the community's decisions regarding the transition process. The college and community are interconnected. Open lectures, free to the public, are frequently offered. Practices and challenges in making the transition are openly discussed. Julia Ponsonby, a vegetarian chef at the college, beautifully ended one of her blogs, *Green Abundance* (2015), on the following note: "Both at Schumacher College and in the local community we are embedded in, we are witnessing the weaving together of the rug of resilience. It may not happen quickly, but if it is well crafted, it is a rug of resilience that will be our magic flying carpet to the future."

Totnes started as a Transition Town in 2006. Since that time, the Transition Town movement has spread to more than eleven hundred towns in forty-three countries (Stites 2013). At the time that Jessica Stites wrote an article about it, in 2013, there were 139 Transition Towns in the United States. On the web, you can find information on how to start a Transition Town. For instance, at TransitionNetwork.org, conference dates, books, films, and other support information is present to help people start their own projects. Additionally, at the Transition U.S. website, blogs on building community and a new "Transition Streets" project have been introduced, where the goal is to bring neighbors together one street at a time, providing them with ideas on how to make simple changes in their households, instilling in them hope instead of fear, and from the ground up, building stronger communities street by street.

Overall, both the Oberlin Project and the Transition Town project are excellent examples of highly successful actions presently aimed at transitioning communities to a land ethic worldview. I have emphasized these two projects because I have firsthand knowledge of them and I find them both inspiring. There are enough other projects, however, to fill a book. In fact, I refer you to a fascinating book by Andree Edwards (2010), *Thriving beyond*

Sustainability, to gain a broader perspective on these other activities. The message I wish to convey to you, however, is that when we think about transitioning to a land ethic worldview, it is not as if we need to start from scratch. Efforts are already well under way, and different paths for beginning the transition are available for people to use in their own communities.

EVERYDAY TRANSITIONS

Seeing projects like the Oberlin Project and Transition Town Totnes is exciting. These are big projects, though, that involve substantial money, leadership, and buy-in from the community. Community meetings must be coordinated and committees created in order to make it all happen. While both projects are trying to be transparent and helpful by providing steps for other communities to follow to make the transition happen, for some the idea may seem to be simply too much.

In another, less complicated way, the transition to a land ethic worldview is happening and can happen very easily. Work conducted by my colleagues and myself shows that all that really needs to happen is for people to be involved with nature in one form or another. For instance, even limited time spent in nature, such as a fifteen-minute walk in the woods, increases individuals' Land Ethic Scale scores or land ethic worldview (Mayer, Frantz, Bruehlman-Senecal, and Dolliver 2009). Our work suggests that being involved in gardening, viewing videos of nature, or viewing nature from inside a building can positively influence a person's land ethic worldview. Activities like these are far less involved than the Oberlin Project and Transition Town movement. They might serve as possible starting points for individuals and groups.

As I mentioned earlier, Oberlin has large town square and an arboretum. On a personal note, I found that when my son was a child, my wife and I could easily spend hours with him playing on the square and in the Arb. We would also go to the local pool and bike ride through the countryside on the bike path. These were very simple, everyday things that individuals and families can do.

Whether to engage in such activities alone or with loved ones is an interesting question that research by Elisabeth Kals, Daniel Schumacher, and Leo Montada (1999) addresses. They investigated how sharing experiences with loved ones influences the impact that exposure to nature has on an individual. What they found is that love of nature was, as expected, predicted by how frequently an individual spent time in nature, in both the past and the present. This finding supports the importance of integrating nature experiences into one's daily routines. Additionally, though, the impact of these experiences was substantially increased by the extent to which they were shared with family members, one's partner, or friends. Thus, while being in nature is important, that the impact of the experience is amplified by having a loved one present is an interesting point to remember.

Their research reminds me of a family outing I took part in when I was a child growing up in Los Angeles. One of my favorite childhood memories involved being in nature with my father at Zuma Beach, just north of L.A. It is a beautiful beach with (in my childhood experience) huge waves. I remember sitting on my father's shoulders as we entered the surf, and my father confidently walking through the face of a wave, which towered over our heads. I recall feeling exhilarated, safe, and marveling at the beauty of the wave as we strode into its icy face. I can still feel the surge of the wave as we passed through it, and still hear the wave crash behind us. My senses were alive, and my love of the ocean and nature was clearly nurtured by this and similar early, shared childhood experiences.

My wife recalls working with her father on his garden. To enrich the soil before planting, they'd go to a nearby horse farm and load manure into the back of a truck. She laughs as she tells the story of how they'd stink up the neighborhood with the manure that sat in the driveway at their house before they incorporated it into the soil. Working in the garden with her father led her to develop a wonderful love of gardening, which she has shared with me and others. It was a gateway experience that led her to develop a deep love of nature and a deep feeling of connection to the natural world.

THE CHILDREN AND NATURE NETWORK

There are also organizations, such as the Children and Nature Network, designed to help people get out into the natural world and connect with nature. This organization was developed in response to the writings of Richard Louv, who popularized the term *nature deficit disorder*. In his book *Last Child in the Woods: Saving Our Children from Nature-Deficit Disorder*, he uses this term not as an official medical diagnosis but as a description of "the human costs of alienation from nature, among them: diminished use of the senses, attention difficulties, and higher rates of physical and emotional illnesses" (2008, p. 34). His work accentuates the many benefits of feeling connected to nature.

His writings have resonated with many people and have inspired, among other groups, the Children and Nature Network. As stated on their website, "The Children & Nature Network is a leading organization in the movement to connect all children, their families, and their communities to nature through innovative ideas, evidence-based resources and tools, broad-based collaboration, and support of grassroots leadership." The website provides a wealth of information on research, programs, and new initiatives. Although primarily aimed at reconnecting children to nature, as their mission statement emphasizes, they are involved in reconnecting families and communities to nature through such projects as greening cities. With grassroots campaigns and clubs in the majority of states and in many countries of the world, this movement is growing and making good use of technology to distribute information and connect people with one another.

In Louv's most recent book, *Vitamin N: The Essential Guide to a Nature Rich Life* (2016), he identifies N, or Nature, as an essential nutrient, one that we all need in order to live a healthy life. In this book he offers five hundred practical activities for newborns, children, youth, and adults, which are aimed at reconnecting people to nature. The activities in this book have the added benefit of "the building of stronger relationships within the family, among friends, and in the community." There is clearly something for everyone in this book.

Starting with the newborn, for instance: "In your child's first months and years, and beyond, go to a park together, spread out a blanket, lie side by side for an hour or more; look up through moving leaves and branches at clouds or moon or stars. Bring water and milk. You may be there for a long time." Or, in order to reduce the possible social isolation of a parent,

> join one of the networks of parents who take their infants and toddlers into nature. "If you have an infant or toddler, consider organizing a neighborhood stroller group that meets for weekly nature walks," suggests the National Audubon Society. Existing networks include the Yahoo group Nature Stroller (Groups.Yahoo.com), which organizes walking and hiking groups for

> families with babies, toddlers, and young children. ToddlerTrails.com provides a directory linking moms, dads, and grandparents to fun "toddler/family" friendly activities, places, and things to do in Orange County, California, and surrounding cities. If such a network doesn't exist in your region, band together with other parents and start one. (Louv 2016, pp. 5–6)

As you can see, this book is an excellent resource.

The activities go on and on. For younger kids, activities include playing in mud puddles, climbing trees, and building tree houses. For older kids, studying constellations and taking a course in basic navigation or orienteering are recommended. Other activities include creating indoor gardens, making a mental map of your bioregion, and even monitoring monarchs. The wide range of activities span different age groups and stretch from inside one's home to the backyard to the local community and beyond. Besides connecting people to nature, nurturing resilience in children and adults is another theme developed in this book. As Louv states, "Falling down is part of a well-balanced childhood. (And for that matter, adulthood.) Children love exploring the dangers of nature—especially if there's a positive adult who helps them feel secure enough to take healthy risks, and if they fall, to learn to stand up again." Overall, this extremely useful book and the additional online resources at the Children and Nature Network only add to the possibilities of activities available to everyone.

URBAN PLANNING

If encountering nature is the surest way to transition to a land ethic worldview, then providing accessible, attractive, and safe green spaces to people living in towns and cities is important. Fortunately, urban planning aimed at incorporating green spaces and parks in urban areas has a long history in the United States and throughout the world. In the United States, New York City's Central Park is a prime example. Developed by Frederick Law Olmstead and Calvert Vaux in 1858 on 778 acres, it has provided urban dwellers a place to escape from the noise and chaotic life of the city. Today, it is one of the most visited urban parks in the United States, with 40 million visitors in 2013 alone.

The heritage of incorporating green spaces in urban centers can be seen in a number of cities in the United States and elsewhere. Cleveland's emerald necklace, a ring of metro parks surrounding this urban area, was initiated in

1917. This set of linked parks now has a beautiful bike path that draws many bike enthusiasts to these parks. With the nearby Cuyahoga National Park, and the Erie Canal towpath that has been converted to a bike path, Cleveland is rich in bike and walking trails. Chicago's lakefront is another prime example of urban development that has incorporated green space within densely populated sections of a city.

The process of greening our towns and cities needs to continue, however. Other activities that Louv encourages people to engage in include creating "wildlife and childlife corridors on private land. Botanical gardens in each city can help create a De-Central Park by declaring that all the green spots of a city from pocket parks to community gardens to green roofs, be viewed as one great park or wildlife corridor. . . . Make sure ordinances and legislation encourage backyard gardens, community gardens, and organic farms" (2016, pp. 231–232). Another activity is to

> Help truly green your cityscape. Push for better urban planning and in developing and redeveloping areas, including tree-planting guidelines, more natural parks, and walkable neighborhoods. Lobby for affordable public transportation, so that urban children and families can easily reach nature areas. The Rails to Trails Conservancy supports former rail lines that have been converted to multi-use trails, for transportation as well as recreation. Urge developers and builders to create green communities, or, better yet, renovate decaying neighborhoods and shopping malls with green oases—urban ecovillages, button parks, and other natural park spaces—that connect children and adults to nature. (Louv 2016, p. 232)

THE ROLE OF CHURCHES

When Lynn White (1967) wrote his highly influential article "The Historical Roots of Our Ecological Crisis," he articulated how religion, especially Christianity, was one main cause for this crisis. This article led early environmentalist to not consider the church as a possible ally in efforts to confront the climate crisis. Years later, Gary Gardner's (2006) book *Inspiring Progress: Religion's Contributions to Sustainable Development* captured the positive role that religion can play, and is playing, in the environmental movement. In the individualistic world where people often find themselves feeling isolated, religions have communities that can take action. Religious leaders also have a moral voice that can inspire followers to act.

Moreover, some Christian denominations have increasingly become focused on the sacredness of nature and how we are all part of the community of creation. In many ways, the image of the body of Christ has been extended to the natural world. In this way the distancing between humans and nature, the fundamental split that White talks about, has been erased. As a consequence, caring for nature is viewed as a moral obligation. Moreover, given Christianity's focus on helping the poor, taking action to confront climate change is a priority because of the moral concern that the poor and powerless are the most vulnerable in life on this new Eaarth. With community, money, and power to make a difference, more and more religious communities singly and together are taking a proenvironmental stance. For example, the United Church of Christ on their environmental ministries web page states, "We are called to go beyond lifestyle adjustment. We are called to spiritual and lifestyle transformation based on justice and reverence for all of God's creatures and creation." This religious orientation nurtures many if not all of the characteristics of the land ethic worldview.

It is somewhat ironic that, although White laid much of the blame for the environmental crisis at the feet of Christianity, at the end of his article he held up Saint Francis as the patron saint of the environmental movement. Today, we find the pope, who has named himself after Saint Francis, presenting an environmental encyclical. Pope Francis's encyclical corresponds very closely to the arguments I've been making in this book. It provides a strong statement about how we are all embedded in the natural world. He argues that our mastery orientation has led us to plunder and harm nature. He links our destruction of nature with our cultural values, especially to a self-centeredness, in which he sees the misuse of nature beginning "when we no longer recognize any higher instance than ourselves, when we see nothing else but ourselves." He goes on to encourage us to look for solutions not only in technology but also "in a change of humanity; otherwise we would be dealing merely with symptoms." Having the moral authority of the pope encourage followers to transition to a new relationship to the natural world is heartening.

Some environmental organizations, too, exhibit a spiritual component. For instance, the ecumenical activity of Greenfaith encourages people to grow spiritually through feeling connected to the earth and to think holistically about environmental problems. Moreover, on their website they assert that "based on beliefs shared by the world's great religions—we believe that protecting the earth is a religious value, and that environmental stewardship is a moral responsibility." The mission of another organization involved in ecumenical work, Interfaith Power and Light, is "to be faithful stewards of Creation by responding to global warming though the promotion of energy conservation, energy efficiency, and renewable energy. This campaign intends to protect earth's ecosystems, safeguard the health of all Creation, and ensure sufficient, sustainable energy for all" (Interfaith Power and Light n.d.).

As you can see, the transition to a land ethic worldview is occurring in many different ways. Many opportunities are available for people to become involved in this transition, and many people have already actively transitioned or are transitioning to this different way of relating to one another and to the natural world. With cooperation, interconnectedness, and an egalitarian spirit, this spreading movement is making a difference in the world.

FINAL THOUGHTS

I am completing this final chapter at a time of great discord and heartbreak in our country. Mass shootings in the United States seem to occur weekly. The tendency to demonize people who are different from us seems only to grow. Instead of coming together to address our troubles, people shout across picket lines or aisles in Congress. Instead of airing differences and finding ways to compromise, people all too often turn dialogue into statements that only polarize us further. The divide between black and white, rich and poor, legal and illegal citizens, Christian and Muslim, and our political leaders seems only to widen.

While writing and thinking about these issues—about life on Eaarth, the divisions among us, and the challenges before us—I started rereading Steinbeck's *The Grapes of Wrath*. The story takes place at the time of the Dust Bowl. At this time farmers were forced to leave their land, and thousands migrated to a place of promise. The divisions between rich and poor were extreme, much as they are today. The story centers on the Joad family and Casy, who serves as Steinbeck's moral voice in the text. You probably know the story, having read it in high school or college. It is about a family living on the edge and about a man, Tom Joad, who comes to see his connection to the land and to others, and his responsibility to help.

Two quotes from this book stand out. In chapter 28, when Tom says good-bye to his mother, he reflects on Casy's wisdom: "Says one time he went out in the wilderness to find his own soul, an' he foun' he didn't have no soul that was his'n. Says he foun' he jus' got a little piece of a great big soul. Says a wilderness ain't no good, 'cause his little piece of a soul wasn't no good 'less it was with the rest, an' was whole."

An interesting aspect of this quote is the remark that a sense of connection to nature by itself wasn't enough. It needed to be combined with a sense of feeling connected to others. "Says a wilderness ain't no good . . . 'less it was with the rest, an' was whole," for the whole includes both nature and other people. Nature by itself is just a part; likewise, people are just a part of the whole. Or, thinking about the Schwartz model one last time: in Schwartz's diagram, self-transcendence includes both egalitarian concern for others and social justice, and the harmony-with-nature component that incorporates unity with and caring for the natural world.

As we've seen in the research presented in chapter 4, when you find people with this sense of feeling connected to nature, you find caring for the environment. And as research in social psychology has found time and again, when people have a sense of interrelatedness to others, they are more likely to help others. Or, as Steinbeck writes after Tom comes to see his interrelatedness and responsibility for others: "Whenever they's a fight so hungry people can eat, I'll be there. Whenever they's a cop beatin' up a guy, I'll be there. If Casy knowed, why, I'll be in the way guys yell when they're mad an'—I'll be in the way kids laugh when they're hungry n' they know supper's ready. An' when our folks eat the stuff they raise an' live in the houses they build—why, I'll be there." Or, in light of today's struggles, whenever an innocent person is being harmed, we all need to "be there," as we need to be there to care for the natural world as well.

If the goal of interconnectedness is to be achieved, then, we need to keep in mind the twofold nature of this goal: connecting people to other people and connecting people to nature. As noted in chapter 4, research has demonstrated that the land ethic worldview does achieve this dual goal: people who feel communality with nature also feel a stronger social cohesion between themselves and others. Additionally, the present chapter emphasizes that the transitions taking place are connecting people to nature and people to one another.

This emphasis on people connecting with people can be seen in the egalitarian spirit and cooperation that forms the bedrock of this movement. The town meetings, the community festivities on the square, the open lectures, the sharing of visions, and the simple gatherings at the open market all contribute to connecting people to people, and people to nature. Louv's five hundred practical ways to connect with nature include many cooperative actions between people. Cooperation breeds a sense of interdependence, and in this way many of Louv's activities create greater harmony between people and between humans and nature. Moreover, activities that are tying people to one another are in all likelihood also creating more resilient communities.

Another point to consider is that in order for this transition to take place, and for it to affect broader aspects of the political life of the nation, it is not necessary for everyone to have first made this transition to a land ethic worldview. Consider the civil rights movement of the 1960s for a moment. Important political changes did not result from this movement because the *entire* nation had come to see the Jim Crow laws and the general discrimination against African Americans in a negative light. In fact, the political scientists Chenoweth and Stephan (2011) point out that only 3.5 percent of a population is needed to topple any government. Certainly, we are not talking about toppling a government, but about exercising political power for legislation to address climate change and human rights.

Lastly, in the big picture, seeing clearly means seeing connections—connections between ourselves and nature, and between ourselves and others. And from this clearer vision will flow thoughts, feelings, and actions that will reorient how we approach the world. From this clearer vision a different path emerges before us. Taking this path involves a new lifestyle. It involves beautiful natural spaces and biophilic architecture that connects people to nature. It involves a feeling of one among many. On this path, where people feel more connected to one another and to nature, they are less likely to be invisible to one another. As we tread, we are more likely to consider how our steps affect the natural world around us, and to leave a smaller carbon footprint behind us. On this path, we will find beauty and hardship as we experience a greater sense of meaning and purpose that surpasses a narrow self-interest. We are no longer bystanders but upstanders, caring for one another and for nature. We are more likely to "be there" to help as we walk on this new Eaarth, with its many challenges. Ultimately, though, it is a path where human systems and natural systems can be in greater harmony with one another, and where there is a greater promise for our children and our children's children. This is a worthy life task to pursue.

REFERENCES

Adler, A. 1956. *The Individual Psychology of Alfred Adler.* Edited by H. L. Ansbacher and R. R. Ansbacher. New York: Harper and Row.

Alicke, M. D. 1985. Global self-evaluation as determined by the desirability and controllability of trait adjectives. *Journal of Personality and Social Psychology* 49:1621–1630.

Alicke, M. D., and Govorun, O. 2005. The better-than-average effect. In *The Self in Social Judgment: Studies in Self and Identity*, edited by M. D. Alicke, D. A. Dunning, and J. I. Krueger, pp. 85–106. New York: Psychology Press..

Amabile, T. M. 1983. *The Social Psychology of Creativity.* New York: Springer.

_____. 1996. *Creativity in Context: Update to the Social Psychology of Creativity.* Boulder, CO: Westview Press.

Audubon Society. 2014. *Audubon's Birds and Climate Change Report: 314 Species on the Brink*. http://climate.audubon.org.

Bandura, A. 1977. Self-efficacy: Toward a unifying theory of behavioral change. *Psychological Review* 84:191–215.

Baron, R. A. 1977. *Human Aggression.* New York: Plenum.

BBC News. 2003. Water hot spots: Ogallala aquifer. Special report. http://news.bbc.co.uk/2/shared/spl/hi/world/03/world_forum/water/html/ogallala_aquifer.stm.

Belch, G. E. 1982. The effects of television commercial repetition on cognitive response and message acceptance. *Journal of Consumer Research* 9:56–65.

Berman, M. G., Jonides, J., and Kaplan, S. 2008. The cognitive benefits of interacting with nature. *Psychological Science* 12:1207–1212.

Brown, J. D. 1986. Evaluations of self and others: Self-enhancement biases in social judgments. *Social Cognition* 4:353–376.

_____. 1998. *The Self.* New York: McGraw-Hill.

Brown, K. W., and Kasser, T. 2005. Are psychological and ecological well-being compatible? The role of values, mindfulness, and lifestyle. *Social Indicators Research* 74:349–368.

Brown, L. 2010. Rising temperatures raise food prices—Heat, drought, and a failed harvest in Russia. Plan B Updates, Earth Policy Institute. www.earth-policy.org.

Bruner, J. 1990. *Acts of Meaning (Four Lectures on Mind and Culture—Jerusalem-Harvard Lectures)*. Cambridge, MA: Harvard University Press.

Brunswik, E. 1939. The conceptual focus of some psychological systems. *Journal of Unified Science* 8:36–49.

———. 1952. *The Conceptual Framework of Psychology*. Vol. 1 of *International Encyclopedia of Unified Science*. Chicago: University of Chicago Press.

Budescu, D. V., Broomell, S., and Por, H.-H. 2009. Improving communication of uncertainty in the reports of the Intergovernmental Panel on Climate Change. *Psychological Science* 20:299–308.

Burke, M. C., and Edell, J. A. 1986. Ad reactions over time: Capturing changes over time. *Journal of Consumer Research* 13:114–118.

Burroughs, J. E., and Rindfleisch, A. 2002. Materialism and well-being: A conflicting values perspective. *Journal of Consumer Research* 29:348–370.

Campbell, T. H., and Kay, A. C. 2014. Solution aversion: On the relation between ideology and motivated disbelief. *Journal of Personality and Social Psychology* 107:809–824.

Capaldi, C. A., Dopko, R. L., and Zelenski, J. M. 2014. The relationship between nature connectedness and happiness: A meta-analysis. *Frontiers of Psychology* 5:1–15.

Carlisle, J., and Smith, E. 2005. Postmaterialism vs. egalitarianism as predictors of energy-related attitudes. *Environmental Politics* 14:527–540.

Catton, B. 1988. *The Civil War*. New York: Fairfax Press.

Chamberlin, S. 2009. *The Transition Timeline: For a Local, Resilient Future*. White River Junction, VT: Chelsea Green Publishing.

Chenoweth, E., and Stephan, M. J. 2011. *Why Civil Resistance Works: The Strategic Logic of Nonviolent Conflict*. New York: Columbia University Press.

Choma, B. L., Hanoch, Y., Gummerum, M., and Hodson, G. 2013. Relations between risk perceptions and socio-political ideology are domain- and ideology-dependent. *Personality and Individual Differences* 54:29–34.

Clayton, S. 2003. Environmental identity: A conceptual and an operational definition. In *Identity and the Natural Environment*, edited by S. Clayton and S. Opotow, pp. 45–65. Cambridge, MA: MIT Press.

Coulson, J., Whitfield, D. H., and Preston, A., eds. 2003. *Keeping Things Whole: Readings in Environmental Science*. Chicago: Great Books Foundation.

Crabtree, Tim. 2015. Blog. schumachercollege.org.uk. Blog discontinued.
Crutzen, P. J., and Stoermer, E. F. 2000. The "Anthropocene." *IGBP Newsletter* 41:17–18.
Dalbert, C. 2001. *The Justice Motive as a Personal Resource: Dealing with Challenges and Critical Life Events*. New York: Plenum.
Darley, J. M., and Latane, B. 1968. Bystander intervention in emergencies: Diffusion of responsibility. *Journal of Personality and Social Psychology* 8:377–383.
Davenport, M. A., and Anderson, D. H. 2005. Getting from sense to place to place-based management: An interpretive investigation of place meanings and perception of landscape change. *Society and Natural Resources* 18:625–641.
Davis, J. L., Green, J. D., and Reed, A. 2009. Interdependence with the environment: Commitment, interconnectedness, and environmental behavior. *Journal of Environmental Psychology* 29:173–180.
Diamond, J. 2005. *Collapse*. New York: Penguin Books.
Diener, E., and Biswas-Diener, R. 2008. *Rethinking Happiness: The Science of Psychological Wealth*. Malden, MA: Blackwell.
Diener, E., Diener, M., and Diener, C. 1995. Factors predicting the subjective well-being of nations. *Journal of Personality and Social Psychology* 69:851–864.
Diener, E., Emmons, R. A., Larsen, R. J., and Griffin, S. 1985. The satisfaction with life scale. *Journal of Personality Assessment* 49:71–75.
Diener, E., and Seligman, M. E. P. 2004. Beyond money: Toward an economy of well-being. *Psychological Science in the Public Interest* 5:1–31.
Dion, K. K., and Dion, K. L. 1993. Individualistic and collectivistic perspectives on gender and the cultural context of love and intimacy. *Journal of Social Issues* 49:53–69.
Dodman, D., Ayers, J., and Hug, S. 2009. "Building resilience." In *State of the World: Into a Warming World*, edited by Linda Starke, pp. 151–168. Washington, DC: Worldwatch Institute.
Doherty, T. J., and Clayton, S. 2011. The psychological impacts of global climate change. *American Psychologist* 4:265–276.
Dunlap, R. E., and Van Liere, K. D. 1978. The new environmental paradigm. *Journal of Environmental Education* 9:10–19.
Dunning, D., Meyerowitz, J. A., and Flolzberg, A. D. 1989. Ambiguity and self-evaluation: The role of idiosyncratic trait definitions in self-serving assessments of ability. *Journal of Personality and Social Psychology* 57:1082–1090.
Durning, A. T. 1995. Are we happy yet? In *Ecopsychology: Restoring the Earth, Healing the Mind*, edited by T. Roszak, M. E. Gomes, and A. D. Kanner. San Francisco: Sierra Club Books.

Dutcher, D. D., Finley, J. C., Luloff, A. E., and Johnson, J. B. 2007. Connectivity with nature as a measure of environmental values. *Environment and Behavior* 39:474–493.

Duval, S., Duval, V. H., with Mayer, F. S. 1983. *Consistency and Cognition: A Theory of Causal Attribution*. Hillsdale, NJ: Lawrence Erlbaum.

Duval, S., Mayer, F. S., Duval, V. H., and DePould, C. 1980. Targets of defensive attribution. Unpublished manuscript, University of Southern California.

Duval, T. S., Silvia, P., and Lalwani, N. 2001. *Self-Awareness and Causal Attribution: A Dual Systems Theory*. New York: Springer.

Easthope, H. 2004. A place called home. *Housing, Theory and Society* 21:128–138.

Edwards, A. E. 2010. *Thriving beyond Sustainability: Pathways to a Resilient Society*. Gabriola Island, BC: New Society.

Einstein, A. 2015. *Bite-Size Einstein: Quotations on Just About Everything from the Greatest Mind of the Twentieth Century*. New York: St. Martin's Press.

Ellison, R. [1952] 1995. *Invisible Man*. New York: Vintage Books.

Erikson, E. [1959] 1980. *Identity and the Life Cycle*. New York: W. W. Norton.

Evans, G. W., and McCoy, J. M. 1998. When buildings don't work: The role of architecture in human health. *Journal of Environmental Psychology* 18:85–94.

Festinger, L. 1954. A theory of social comparison processes. *Human Relations* 7:117–140.

———. 1957. *A Theory of Cognitive Dissonance*. Stanford, CA: Stanford University Press.

Festinger, L., Riecken, H. W., and Schachter, S. 1956. *When Prophecy Fails*. Minneapolis: University of Minnesota Press.

Feygina, I., Jost, J. T., and Goldsmith, R. E. 2010. System justification, the denial of global warming, and the possibility of "system-sanctioned change." *Personality and Social Psychology Bulletin* 36:326–338.

Fishwick, L, and Vining, J. 1992. Toward a phenomenology of recreation place. *Journal of Environmental Psychology* 12:57–63.

Fiske, S. T. 2012. *Social Beings: Core Motives in Social Psychology*. Hoboken, NJ: Wiley.

Frantz, C. M., and Mayer, F. S. 2005a. Connectedness to nature as a source of well-being for humans. Paper presented at the Positive Psychology Summer Institute, Philadelphia.

———. 2005b. Feeling connected to nature: Benefits for self and the environment. Paper presented at the Interdisciplinary Conference on Science and Culture: Eco-Problems/Eco-Solutions, Frankfort, KY.

———. 2009. Why are we failing to take action against the emergency of climate change? *Analysis of Social Issues and Public Policy* 9:205–222.

———. 2013. The importance of connection to nature in assessing environmental education programs. *Studies in Educational Evaluation* 41:85–89. http://dx.doi.org/10.1016/j.stueduc.2013.10.001.

Frantz, C. M., Mayer, F. S., Norton, C., and Rock, M. 2005. There is no "I" in nature: The influence of self-awareness on connectedness to nature. *Journal of Environmental Psychology* 25:427–436.

Frantz, C. M., Mayer, F. S., Petersen, J. and Shamin, M. R. 2013. What best predicts resource conservation behavior? The case for connectedness to nature. Unpublished manuscript.

Freud, S. [1927] 1989. *The Future of an Illusion*. New York: W. W. Norton.

Fritsche, I., Cohrs, J. C., Kessler, T., and Bauer, J. 2012. Global warming is breeding social conflict: The subtle impact of climate change threat on authoritarian tendencies. *Journal of Environmental Psychology* 1:1–10.

Fromm, E. 1955. *The Sane Society*. London: Routledge Classics.

———. 1964. *The Heart of Man*. Harper and Row.

Frumkin, H. 2001. Beyond toxicity: Human health and the natural environment. *American Journal of Preventive Medicine* 20:234–240.

Fry, D. P. 2007. *Beyond War: The Human Potential for Peace*. Oxford: Oxford University Press.

Gardner, G. T. 2006. *Inspiring Progress: Religions' Contributions to Sustainable Development*. New York: W. W. Norton.

Geiger, N., and Swim, J. K. 2016. Climate of silence: Pluralistic ignorance as a barrier to climate change discussion. *Journal of Environmental Psychology* 47:79–90.

Gifford, R. 2011. The dragons of inaction: Psychological barriers that limit climate change mitigation and adaptation. *American Psychologist* 4:290–302.

———. 2014. Environmental psychology matters. *Annual Review of Psychology* 65:541–579.

Gifford, R., Scannell, L., Kormos, C., Smolova, L., Biel, A., Boncu, S., Corral, V., Guntherf, H., Hanyu, K., Hine, D., et al. 2009. Temporal pessimism and spatial optimism in environmental assessments: An 18-nation study. *Journal of Environmental Psychology* 29:1–12.

Gilligan, C. 1982. *In a Different Voice: Psychological Theory and Women's Development*. Cambridge, MA: Harvard University Press.

Goldenberg, S. 2013. Secret funding helped build vast network of climate denial thinktanks. *The Guardian*, February 14.

Gould, S. J. 1996. *The Mismeasure of Man*. New York: W. W. Norton.

Graham, J., Haidt, J., and Nosek, B. A. 2009. Liberals and conservatives rely on different sets of moral foundations. *Journal of Personality and Social Psychology* 5:1029.

Gromet, D. M., Kunreuther, H., and Larrick, R. P. 2013. Political ideology affects energy-efficiency attitudes and choices. *Proceedings of the National Academy of Sciences* 110 (23): 9314–9319.

Grossman, D. 2009. *On Killing: The Psychological Cost of Learning to Kill in War and Society*. New York: Back Bay Books.

The Guardian. 2009. Global warming could create 150 million "climate refugees" by 2050. November 11.

Haidt, J., and Graham, J. 2007. When morality opposes justice: Conservatives have moral intuitions that liberals may not recognize. *Social Justice Research* 1:98–116.

Hammitt, W.E., Backlund, E.A., and Bixler, R.D. 2006. Place bonding for recreation places: Conceptual and empirical development. *Leisure Studies* 25:17–41.

Hansen, J.C. 2010. *Storms of My Grandchildren: The Truth about the Coming Climate Catastrophe and Our Last Chance to Save Humanity*. New York: Bloomsbury USA.

Haring-Hidore, M., Stock, W.A., Okun, M.A., and Witter, R.A. 1985. Marital status and subjective well-being: A research synthesis. *Journal of Marriage and the Family* 47:947–953.

Hatfield, J., and Job, R.F.S. 2001. Optimism bias about environmental degradation: The role of the range of impact of precautions. *Journal of Environmental Psychology* 21:17–30.

Hawcroft, L.J., and Milfont, T.L. 2010. The use (and abuse) of the new environmental scale over the last 30 years: A meta-analysis. *Journal of Environmental Psychology* 30:143–158.

Hedlund-de Witt, A., Boer, J. de, and Boersema, J.J. 2014. Exploring inner and outer worlds: A quantitative study of worldviews, environmental attitudes, and sustainable lifestyles. *Journal of Environmental Psychology* 37:40–54.

Heider, F. 1944. Social perception and phenomenal causality. *Psychological Review* 51:358–374.

Heine, S.J. 2015. *Cultural Psychology*. 3rd ed. New York: W.W. Norton.

Hinkle Charitable Foundation. 2006. *Report 5: How Do We Contribute Individually to Global Warming?* Hinkle Charitable Foundation. www.thehcf.org/report-5-how-do-we-contribute-individually-to-global-warming.

Hoffarth, M.R., and Hodson, G. 2016. Green on the outside, red on the inside: Perceived environmentalist threat as a factor explaining political polarization of climate change. *Journal of Environmental Psychology* 45:40–49.

Hofstadter, R. 1963. *Anti-intellectualism in American Life*. New York: Vintage.

Hofstede, G. 1980. *Culture's Consequences: International Differences in Work-Related Values*. Beverly Hills, CA: Sage.

Hopkins, R. 2008. *The Transition Town Handbook: From Oil Dependency to Local Resilience*. White River Junction, VT: Chelsea Green.

———. 2011. *The Transition Companion: Making Your Community More Resilient in Uncertain Times*. White River Junction, VT: Chelsea Green.

Horgan, J. 2012. *The End of War*. San Francisco: McSweeney's Books.

Howell, A. J., Dopko, R. L., Passmore, H.-A., and Buro, K. 2011. Nature connectedness: Associations with well-being and mindfulness. *Personality and Individual Differences* 51:166–171.

Interfaith Power and Light. N.d. Mission statement. interfaithpowerandlight.org.

Intergovernmental Panel on Climate Change (IPCC). 2007. *Climate Change 2007: The Physical Science Basis; Contribution of Working Group 1 to the Fourth Assessment Report of the Intergovernmental Panel on Climate Change*. Cambridge: Cambridge University Press.

_____. 2014. *Climate Change 2013: The Physical Science Basis; Contribution of Working Group 1 to the Fifth Assessment Report on the Intergovernmental Panel on Climate Change*. Cambridge: Cambridge University Press.

Jacoby, S. 2009. *The Age of American Unreason*. New York: Vintage.

Jacques, R. J., Dunlap, R. E., and Freeman, M. 2008. The organization of denial: Conservative think tanks and environmental skepticism. *Environmental Politics* 17:349–385.

Jorgensen, B. S., and Stedman, R. C. 2001. Sense of place as an attitude: Lakeshore owners' attitudes toward their properties. *Journal of Environmental Psychology* 21:233–248.

_____. 2006. A comparative analysis of predictors of sense of place dimensions: Attachment to, dependence on, and identification with lakeshore properties. *Journal of Environmental Management* 79:316–327.

Jost, J. T., and Benaji, M. R. 1994. The role of stereotypes in system-justification and the production of false consciousness. *British Journal of Social Psychology* 33:1–27.

Jost, J. T., Nosek, B. A., and Gosling, S. D. 2008. Ideology: Its resurgence in social, personality, and political psychology. *Perspectives on Psychological Science* 3:126–136.

Jugert, P., Greenway, K. H., Barth, M., Buchner, R., Eisentraut, S., and Fritsche, I. 2016. Collective efficacy increases pro-environmental intentions through increasing self-efficacy. *Journal of Environmental Psychology* 48:12–23.

Kahn, B. 2016. Climate change is coming for your maple syrup. Climate Central. March 28. www.climatecentral.org/news/climate-change-maple-syrup-20178.

Kals, E., Schumacher, D., and Montada, L. 1999. Emotional affinity toward nature as a motivational basis to protect nature. *Environment and Behavior* 31:178–202.

Kaltenborn, B. P., and Bjerke, T. 2002. Associations between landscape preferences and place attachment: A study in Roros, Southern Norway. *Landscape Research* 27:381–396.

Kamitsis, I., and Francis, A. J. P. 2013. Spirituality mediates the relationship between engagement with nature and psychological well-being. *Journal of Environmental Psychology* 36:136–143.

Kaplan, R., and Kaplan, S. 1989. *The Experience of Nature: A Psychological Perspective*. New York: Cambridge University Press.

Kaplan, S., and Kaplan, R. 2003. Health, supportive environments, and the reasonable person model. *American Journal of Public Health* 93:1484–1489.

Kasser, T., Cohn, S., Kanner, A. D., Ryan, R. M. 2007. Some costs of American corporate capitalism: A psychological exploration of value and goal conflicts. *Psychological Inquiry* 1:1–22.

Kauffman, J., ed. 2002. *Loss of the Assumptive World: A Theory of Traumatic Loss*. Series in Trauma and Loss. New York: Brunner-Routledge.

Keller, J. 2015. What makes Americans so optimistic? *The Atlantic*. March 25.

Kellert, S. R., and Wilson, E. O., eds. 1993. *The Biophilia Hypothesis*. Washington, DC: Island Press.

Kelly, G., and Hosking, K. 2008. Nonpermanent residents, place attachment and "sea change" communities. *Environment and Behavior* 40:575–594.

Kenrick, D. T., Neuberg, S. L., and Cialdini, R. B. 2015. *Social Psychology*. Boston: Pearson.

Kohlberg, L. 1969. *Stages in the Development of Moral Thought and Action*. New York: Holt, Rhinehart, and Winston.

Kohn, A. 1999. *Punished by Rewards: The Trouble with Gold Stars, Incentive Plans, A's, Praise, and Other Bribes*. Boston: Mariner Books.

Krauthammer, C. 2009. Charles Krauthammer on the new socialism. *Washington Post*, December 11.

Kuo, F. E., and Sullivan, W. C. 2001. Aggression and violence in the inner city: Effects of environment via mental fatigue. *Environment and Behavior* 4:543–571.

Kusenback, M. 2008. A hierarchy of urban communities: Observations on the nested character of place. *City and Community* 7:225–249.

Kyle, G., Graefe, A., Manning, R., and Bacon, J. 2003. An examination of the relationship between leisure activity involvement and place attachment among hikers along the Appalachian Trail. *Journal of Leisure Research* 35:249–273.

Laczko, L. S. 2005. National and local attachments in a changing world system: Evidence from an international survey. *International Review of Sociology* 15:517–528.

Larsen, J. 2006. Setting the record straight: More than 52,000 Europeans died from heat in summer of 2003. Plan B Updates, Earth Policy Institute. www.earth-policy.org.

Latane, B., and Darley, J. M. 1969. Bystander apathy. *American Scientist* 57:244–268.

———. 1970. *The Unresponsive Bystander: Why Doesn't He Help?* New York: Appleton-Crofts.

Leiserowitz, A. 2006. Climate change risk perception and policy preferences: The role of affect, imagery, and values. *Climate Change* 77:45–72.

Leiserowitz, A., Maibach, E., Roser-Renouf, C., Feinberg, G., and Rosenthal, S. 2015. *Climate Change in the American Mind: March 2015*. New Haven, CT: Yale University and George Mason University Yale Project on Climate Change Communication.

Leong, L. Y. C., Fischer, R., and McClure, J. 2015. Are nature lovers more innovative? The relationship between connectedness with nature and cognitive styles. *Journal of Environmental Psychology* 40:57–63.

Leopold, A. 1949. *A Sand County Almanac: With Essays on Conservation from Round River*. New York: Ballantine.

Lerner, M. J. 1980. *The Belief in a Just World: A Fundamental Delusion*. New York: Plenum.

Lerner, M. J., and Montada, L. 1998. An overview: Advances in belief in a just world theory and methods. In *Responses to Victimizations and Belief in a Just World*, ed. Leo Montada and M. J. Lerner, 1–7. New York: Plenum Press.

Lewicka, M. 2008. Place attachment, place identity and place memory: Restoring the forgotten city past. *Journal of Environmental Psychology* 28:209–231.

———. 2010. What makes neighborhood different from home and city? Effects of place scale on place attachment. *Journal of Environmental Psychology* 30:35–51.

———. 2011. Place attachment: How far have we come in the last 40 years? *Journal of Environmental Psychology* 31:207–230.

Lewin, K. 1936. *Principles of Topological Psychology*. New York: McGraw-Hill.

Lezak, S. B., and Thibodeau, P. H. 2016. Systems thinking and environmental concern. *Journal of Environmental Psychology* 46:143–153.

Lipkus, I. M., Dalbert, C., and Siegler, I. C. 1996. The importance of distinguishing the belief in a just world for self versus for others: Implications for psychological well-being. *Personality and Social Psychology Bulletin* 22 (7): 666–677.

Lorenzoni, I., Nicholson-Cole, S., and Whitmarsh, L. 2007. Barriers perceived to engaging with climate change among the UK public and their policy implications. *Global Environmental Change* 17:445–459.

Louv, R. 2008. *Last Child in the Woods: Saving Our Children from Nature-Deficit Disorder*. Chapel Hill, NC: Algonquin Books.

———. 2016. *Vitamin N: The Essential Guide to a Nature-Rich Life*. Chapel Hill, NC: Algonquin Books.

Lyubomirsky, S., and Lepper, H. 1999. A measure of subjective happiness: Preliminary reliability and construct validation. *Social Indicators Research* 46:137–155.

Markham, E. 1936. The right kind of people. In *The Best Loved Poems of the American People*, edited by H. Felleman. New York: Doubleday.

Markus, H. R., Kitayama, S., and Heiman, R. 1996. Culture and "basic" psychological principles. In *Social Psychology: Handbook of Basic Principles*, edited by E. Higgins and A. W. Kruglanski, pp. 857–913. New York: Guilford Press.

Marshall, S. L. A. 1947. *Men against Fire*. New York: William Morrow.

Masci, D. 2014. 5 facts about evolution and religion. PEW Research Center. www.pewresearch.org/fact-tank/2014/10/30/5-facts-about-evolution-and-religion/

Maslow, A. H. 1954. *Motivation and Personality*. New York: Harper and Row.

Matsumoto, D., ed. 2001. *The Handbook of Culture and Psychology*. Oxford: Oxford University Press.

Matsumoto, D., and Juang, L. 2012. *Culture and Psychology*. 5th ed. Belmont, CA: Wadsworth.

Mayer, F. S., Duval, S., and Duval, V. H. 1980. An attributional analysis of commitment. *Journal of Personality and Social Psychology* 39:1072–1808.

Mayer, F. S., Duval, S., Holtz, R., and Bowman, C. 1985. Self-focus, helping request salience, felt responsibility, and helping behavior. *Personality and Social Psychology Bulletin* 4:133–144.

Mayer, F. S., and Frantz, C. M. 2004. The Connectedness to Nature Scale: A measure of individuals' feeling in community with nature. *Journal of Environmental Psychology* 24:503–515.

Mayer, F. S., Frantz, C. M., Bruehlman-Senecal, E., and Dolliver, K. 2009. Why is nature beneficial?: The role of connectedness to nature. *Environment and Behavior* 5:607–643.

Mayer, J. 2017. *Dark Money: The Hidden History of the Billionaires behind the Rise of the Radical Right*. New York: Anchor.

McDonald, R. I., Chai, H. Y., and Newell, B. R. 2015. Personal experience and "psychological distance" of climate change: An integrative review. *Journal of Environmental Psychology* 44:109–118.

McIntee, M. 2015. NASA predicts 20–40 year megadroughts in US because of man-made climate change. February 12. *The Uptake*. http://theuptake.org/2015/02/12/nasa-predicts-20-40-year-megadroughts-in-us-because-of-man-made-climate-change/.

McKibben, B. 2010. Eaarth: *Making a Life on a Tough New Planet*. New York: Times Books.

———. 2012a. *The Global Warming Reader: A Century of Writing about Climate Change.* New York: Penguin Books.

———. 2012b. Global warming's terrifying new math. *Rolling Stone*, July 24.

McPherson, J. M. 2003. *Battle Cry of Freedom: The Civil War Era*. New York: Oxford University Press.

Megerian, C., Stevens, M., and Boxall, B. 2015. Brown orders California's first mandatory water restrictions: "It's a different world." *Los Angeles Times*, April 1.

Messick, D. M., Bloom, S., Boldizar, J. P., and Samuelson, C. D. 1985. Why we are fairer than others. *Journal of Experimental Social Psychology* 21:480–500.

Michotte, A. E. 1963. *The Perception of Causality*. New York: Basic Books.

Milfont, T. L., and Duckitt, J. 2010. The environmental inventory: A valid and reliable measure to assess the structure of environmental attitudes. *Journal of Environmental Psychology* 30:80–94.

Milgram, S. 1975. *Obedience to Authority*. New York: Harper Colophon.

Mooney, C. 2015. Want to get conservatives to save energy? Stop the environmental preaching. *Washington Post*, February 12.

Murray, S. L., and Holmes, J. G. 1993. Seeing virtues in faults: Negativity and the transformation of interpersonal narratives in close relationships. *Journal of Personality and Social Psychology* 65:707–722.

———. 1997. A leap of faith? Positive illusions in romantic relationships. *Personality and Social Psychology Bulletin* 23:586–604.

Myers, D. G. 2000. The funds, friends, and faith of happy people. *American Psychologist* 55: 56–67.

Neff, K. D. 2008. Self compassion: Moving beyond the pitfalls of a separate self-concept. In *Transcending Self-Interest: Psychological Explorations of the Quiet Ego*, edited by J. Bauer and H. A. Wayment, pp. 95–105. Washington, DC: APA Books.

New York Times. 2016. What could disappear. *Sunday Review*, April 24.

Nisbet, E. K., and Zelenski, J. M. 2011. Underestimating nearby nature: Affective forecasting errors obscure the happy path to sustainability. *Psychological Science* 22:1101–1106.

Nisbet, E. K., Zelenski, J. M., and Murphy, S. A. 2009. The nature relatedness scale: Linking individuals' connection with nature to environmental concern and behavior. *Environment and Behavior* 41:715–740.

Nisbett, R. E. 2004. *The Geography of Thought: How Asians and Westerners Think Differently . . . and Why*. New York: Free Press.

O'Connor, R. E., Bord, R. J., and Fisher, A. 1998. Rating threat mitigators: Faith in experts, governments and individuals themselves to create a safer world. *Risk Analysis* 18:547–556.

Oishi, S., Lun, J., and Sherman, G. D. 2007. Residential mobility, self-concept, and positive affect in social interactions. *Journal of Personality and Social Psychology* 1:131–141.

Opotow, S. 2005. Hate, conflict, and moral exclusion. In *The Psychology of Hate*, edited by R. J. Sternberg. Washington, DC: American Psychological Association.

Ornstein, R., and Ehrlich, P. 1989. *New World, New Mind: Moving toward Conscious Evolution*. New York: Touchstone.

———. 2000. *New World New Mind*. Cambridge, MA: Malor Books.

Orr, D. W. 1994. *Earth in Mind*. Washington, DC: Island Press.

———. 2007. Hope in a hotter time. In *Hope Is an Imperative: The Essential Davis Orr*, by D. W. Orr. Washington, DC: Island Press, 2011.

———. 2009. *Down to the Wire: Confronting Climate Collapse*. Oxford: Oxford University Press.

———. 2010. Long tails and ethics: Thinking about the unthinkable. In *Hope Is an Imperative: The Essential Davis Orr*, by D. W. Orr. Washington, DC: Island Press, 2011.

———. 2014. The Oberlin project. *Solutions* 5 (May): 7–12.

Oskamp, S. 2000. Psychological contributions to achieving an ecologically sustainable future for humanity. *Journal of Social Issues* 3:373–390.

Oyserman, D., and Lee, S. W.-S. 2010. Priming culture: Culture as situated cognition. In *Handbook of Cultural Psychology*, edited by S. Kitayama and D. Cohen. New York: Guilford Press.

Pahl, S., Harris, P. R., Todd, H. A., and Rutter, D. R. 2005. Comparative optimism for environmental risks. *Journal of Environmental Psychology* 25:1–11.

Perkins, H. E. 2010. Measuring love and care for nature. *Journal of Environmental Psychology* 30:455–463.

Pettigrew, T. F., and Tropp, L. R. 2011. *When Groups Meet: The Dynamics of Intergroup Contact*. New York: Psychology Press.

Ponsonby, Julia. 2015. *Green Abundance*, July 15. schumachercollege.org.uk. Blog discontinued.

Poon, K.-T., Teng, F., Chow, J. T., and Chen, Z. 2015. Desiring to connect to nature: The effect of ostracism on ecological behavior. *Journal of Environmental Psychology* 42:116–122.

Portinga, W., Steg, L., and Vlek, C. 2002. Environmental concern and preference for energy-saving measures. *Environment and Behavior* 34:455–478.

Prentice, D. A., and Miller, D. T. 1993. Pluralistic ignorance and alcohol use on campus: Some consequences of misperceiving the social norm. *Journal of Personality and Social Psychology* 2:243.

Price, J. C., Walker, I. A., and Boschetti, F. 2014. Measuring cultural values and beliefs about environment to identify their role in climate change responses. *Journal of Environmental Psychology* 37:8–20.

Putnam, R. D. 2001. *Bowling Alone: The Collapse and Revival of American Community*. New York: Touchstone Books.

Reicher, S., Hopkins, N., and Harrison, K. 2006. Social identity and spatial behaviour: The relationship between national category salience, the sense of home, and labour mobility across national boundaries. *Political Psychology* 27:247–263.

Reser, J. P., and Swim, J. K. 2011. Adapting to and coping with the threat and impacts of climate change. *American Psychologist* 4:277–289.

Rice, D. 2014. California's 100-year drought: Megadroughts a threat to civilization. *USA Today*, September 3. See also Union of Concerned Scientists, *Causes of Drought: What's the Climate Connection?* which can be obtained online.

Ritter, C., Benson, D. E., and Snyder, C. 1990. Belief in a just world and depression. *Sociological Perspective* 25:235–252.

Rogers, M., Miller, N., Mayer, F. S., and Duval, S. 1982. Personal responsibility and salience of the request for help: Determinants of the relation between negative affect and helping behavior. *Journal of Personality and Social Psychology* 5:956–970.

Ross, L., and Nisbett, R. E. 1991. *The Person and the Situation: Perspectives in Social Psychology*. New York: McGraw-Hill.

Ross, M. 1981. Self-centered biases in attributional responsibility: Antecedents and consequences. In *Social cognition: The Ontario symposium*, edited by E. T. Higgins, C. P. Herman, and M. P. Zanna, pp. 305–321. Hillsdale, NJ: Erlbaum.

Rossen, I. L., Dunlop, P. D., and Lawrence, C. M. 2015. The desire to maintain the social order and the right to economic freedom: Two distinct moral pathways to climate change scepticism. *Journal of Environmental Psychology* 42:42–47.

Roszak, T. 1992. *The Voice of the Earth*. Grand Rapids, MI: Phanes Press.

Sallenger, A. H., Doran, K. S., and Howd, P. A. 2012. Hotspot of accelerated sea-level rise on the Atlantic coast of North America. *Nature Climate Change* 2. nature.com.

Savitsky, K., Boven, L. V., Epley, N., and Wight, W. M. 2005. The unpacking effect in allocations of responsibility for group tasks. *Journal of Experimental Social Psychology* 41:447–457.

Schultz, P. W. 2001. Assessing the structure of environmental concern: Concern for the self, other people, and the biosphere. *Journal of Environmental Psychology* 21:327–339.

Schumacher College. N.d. About the college. www.schumachercollege.org.uk/about.

Schwartz, S.H. 1992. Universals in the content and structure of values: Theory and empirical tests in 20 countries. In *Advances in experimental social psychology*, edited by M. Zanna, vol. 25, pp. 1–65. New York: Academic Press.

———. 1994. Beyond individualism/collectivism: New cultural dimensions of values. In *Individualism and Collectivism: Theory, Method, and Applications*, edited by U. Kim, H.C. Triandis, C. Kagitcibasi, S.-C. Choi, and G. Yoon. Thousand Oaks, CA: Sage.

———. 1996. Value priorities and behavior: Applying a theory of integrated value systems. In *The Psychology of Values: The Ontario Symposium*, edited by C. Seligman, J.M. Olson, and M.P. Zanna, vol. 8, pp. 1–24. Hillsdale, NJ: Erlbaum.

———. 2007. Cultural and individual value correlates of capitalism: A comparative analysis. *Psychological Inquiry* 1:52–57.

———. 2008. Structural equivalence of the values domain across cultures: Distinguishing sampling fluctuations from meaningful variation. *Journal of Cross-Cultural Psychology* 4:345–365.

Sedikides, C., and Strube, M.J. 1997. Self-evaluation: To thine own self be good, to thine own self be sure, to thine own self be true, and to thine own self be better. In *Advances in Experimental Social Psychology*, edited by M. Zanna, vol. 29, pp. 209–269. New York: Academic Press.

Segall, M.H., Campbell, D.T., Herskovits, M.J. 1966. *The Influence of Culture on Visual Perception*. Oxford, UK: Bobbs-Merrill.

Seiferlein, Katherine E. 2004. *Annual Energy Review, 2003*. Washington, DC: USDOE Energy Information Administration. www.osti.gov/scitech/servlets/purl/1184624.

Seligman, M.E.P. 2011. *Flourish: A Visionary New Understanding of Happiness and Well-Being*. New York: Simon and Schuster.

Shakespeare, W. 1997. *King Lear*. Arden Shakespeare, 3rd series. London: Bloomsbury Arden Shakespeare.

Smaldone, D. 2006. The role of time in place attachment. In *Proceedings of the 2006 Northeastern Recreation Research Symposium*. Newtown Square, PA: Department of Agriculture, Forest Service, Northern Research Station.

Solomon, S., Greenberg, J., Schimel, J., Arndt, J., and Pyszczynski, T. 2004. Human awareness of mortality and the evolution of culture. In *The Psychological Foundations of Culture*, edited by M. Schaller and C. Crandall, pp. 15–40. New York: Erlbaum.

Sowell, T. 2001. *The Quest for Cosmic Justice*. New York: Simon and Schuster.

Stedman, R.C. 2003. Is it really just a social construction? The contribution of the physical environment to sense of place. *Society of Natural Resources* 16:671–685.

Steinbeck, J. [1939] 2006. *The Grapes of Wrath*. New York: Penguin Classics.
Sternberg, R. J. 2012. "The intelligence of nations: Smart but not wise—a comment on Hunt 2012." *Psychological Science* 2:187–189.
Stites, J. 2013. Is your town in transition? *In These Times*, May 20.
Stoessinger, J. G. 2010. *Why Nations Go to War*. Boston: Wadsworth, Cengage Learning.
Stollberg, J., Fritsche, I., and Baecker, A. 2015. Striving for group agency: Threat to personal control increases the attractiveness of agentic groups. *Frontiers in Psychology* (May 27): 6. http://dx.doi.org/10.3389/fpsyg.2015.00649.
Strathman, A., Gleicher, F., Boninger, D. S., and Edwards, C. S. 1994. The consideration of future consequences: Weighing immediate and distant outcomes of behavior. *Journal of Personality and Social Psychology* 66:742–752.
Svenson, O. 1981. Are we all less risky and more skillful than our fellow drivers? *Acta Psychologica* 47:143–148.
Swim, J. K., Stern, P. C., Doherty, T. J., Clayton, S., Reser, J. P., Weber, E. U., Gifford, R., and Howard, G. S. 2011. Psychology's contributions to understanding and addressing global climate change. *American Psychologist* 4:241–250.
Tajfel, H., and Turner, J. C. 1986. The social identity theory of intergroup behavior. In *Psychology of Intergroup Relations*, edited by S. Worchel and W. Austin, pp. 7–24. Chicago: Nelson-Hall.
Tam, K.-P. 2013. Concepts and measures related to connection to nature: Similarities and differences. *Journal of Environmental Psychology* 34:64–78.
Taylor, S. E., and Brown, J. D. 1988. Illusion and well-being: A social psychological perspective on mental health. *Psychological Bulletin* 103:193–210.
Teixeira, B. 1999. Nonviolence theory and practice. In *Encyclopedia of Violence, Peace, and Conflict*, edited by L. Kurtz, pp. 555–565. New York: Academic Press.
Triandis, H. C. 1996. Cultural syndromes. *American Psychologist* 4:407–415.
Tuan, Y.-F. 1975. Place: An experiential perspective. *Geographical Review* 65:151–165.
Twenge, J. M., and Campbell, W. K. 2009. *The Narcissism Epidemic: Living in the Age of Entitlement*. New York: Free Press.
Twenge, J. M., Campbell, W. K., and Gentile, B. 2012. Changes in pronoun use in American books and the rise of individualism, 1960–2008. *Journal of Cross-Cultural Psychology* 3:406–415.
United Nations. 2012. *World Urbanization Prospects, the 2011 Revision*. New York: United Nations.
van der Linden, S. L., Leiserowitz, A. A., Feinberg, G. D., and Maibach, E. W. 2015. The scientific consensus on climate change as a gateway belief: Experimental evidence. *PLoS ONE* 10:1–8.

Van Lange, P. A. M., and Rusbult, C. E. 1995. My relationship is better than—and not as bad as—yours is: The perception of superiority in close relationships. *Personality and Social Psychology Bulletin* 21:32–44.

Wei, R., Lo, V.-H., and Lu, H.-Y. 2007. Reconsidering the relationship between the third-person perception and optimistic bias. *Communication Research* 34:665–684.

Weinstein, N. D. 1980. Unrealistic optimism about future life events. *Journal of Personality and Social Psychology* 39:806–820.

Weinstein, N. D., Klotz, M. L., and Sandman, P. M. 1988. Optimistic biases in public perceptions of the risks of radon. *American Journal of Public Health* 78:796–800.

White, L. 1967. The historical roots of our ecologic crisis. *Science* 155:1203–1207.

Wight, J. 2012. *Titanic* and global warming. Precarious Planet. https://jameswight.wordpress.com/2012/04/15/titanic-global-warming/.

Williams, D. R., and Van Patten, S. 1998. Back to the future? Tourism, place, and sustainability. In *Sustainability and Development: On the Future of Small Society in a Dynamic Economy*, edited by L. Anderson, and T. Blom, pp. 359–369. Karlstad, Sweden: University of Karlstad.

Wilson, E. O. 1984. *Biophilia*. Cambridge: MA: Harvard University Press.

Witter, R., Okun, M., Stock, W. A., and Haring, M. 1984. Education and subjective well-being: A meta-analysis. *Education Evaluation and Policy Analysis* 6:165–173.

Woodley, R. S. 2012. *Shalom and the Community of Creation: An Indigenous Vision*. Grand Rapids, MI: Eerdmans.

World Commission on Environment and Development (WCED). 1987. *Our Common Future*. Oxford. Oxford University Press.

Wright, R. 2004. *A Short History of Progress*. Edinburgh, UK: Canongate.

Zelenski, J. M., and Nisbet, E. K. 2014. Happiness and feeling connected: The distinct role of nature relatedness. *Environment and Behavior* 46:3–23.

INDEX

"fig." refers to figures

action abilities (stage 5), 68, 97–101
actions, 151–73; barriers and, 67, 72; beliefs and, 40; carbon (CO2) emissions and, 8, 155; certainties and, xxiii; the Children and Nature Network, 165–67; Christianity/churches/religion and, 168–70; clarity and, 151; climate change and, 3; collective actions, 24–25, 97, 105; communities and, 168; consumer practices and, 134; cosmetic v curative and, 8–9; cost/benefit analyses and, 68; cultural lenses/schemas/worldviews and, xx, 18–19, 39; cultures and, 17–18; dis/connection and, 68–69, 143; distortions and, 1, 20; Easter Island and, 20; environmental dashboards and, 155; errors/misjudgments and, xxiii; everyday transitions, 164–65; goals and, 151; holistic thinking and, 143; hope and, 107; hot tub example and, 74–75; ignorance and, 81; interconnectedness and, 172; land ethic worldview and, 132, 149, 151–73; mistrust and, 101; mitigation and, 11; natural world and, 44, 81, 113, 115*fig.*, 120–21; the Oberlin Project and, 152–59, 160, 163, 164; obstacles and, xxiii, 68, 104; personal responsibilities (stage 3) and, 68–70; psychologists and, 12; science and, xxiv, 109; temporal risks and, 102; tokenism and, 104; Transition Town Totnes, 160–64; urban planning, xvii, 167–68; water and, 6, 8; worldviews and, xix, 24–25, 65, 106. *See also* carbon (CO2) emissions; failure of actions; political activism
Adam J. Lewis Environmental Studies Center, 153
adaptations, 11, 34*fig.*, 73*fig.*, 90, 145*fig.*, 146
Affect Balance Scale, 139
Affective Autonomy (Schwartz model), 46–47, 112*fig.*
affects, 10*fig.*, 72–73, 134, 140*fig.*, 145*fig.*
Afghanistan, 6
African Americans, 32, 152, 172
Age of American Unreason, The (Jacoby), 85
aggression, 35, 37–38. *See also* conflicts; intergroup conflicts
agriculture, xvii, 4, 5–6, 20, 153–55, 158. *See also* organic food
airport example, 97–98
alternative worldviews, xxiii, 11, 21, 22, 26, 30, 109. *See also* land ethic worldview

Amabile, Teresa, 59–60, 99–100
American corporate capitalism (ACC), 22–23
American Psychological Association (APA), 10
analytic thinking, 40, 61–62, 97*fig.*, 100, 106, 113, 152
ancient brains, xvi, 75*fig.*, 76–77, 80
ancient worlds, 27, 37, 84–85
anger, 91–92. *See also* intergroup conflicts
Animate Earth (Harding), 161
Anthropocene period, 4, 149
anti-intellectualism, 84–85, 88, 89, 105
Anti-intellectualism in American Life (Hofstadter), 85
aquifers, 5, 20–21
Asia, 6–7
aspirations, 102*fig.*, 103–4
"Assessing the Structure of Environmental Concern: Concern for the Self, Other People, and the Biosphere" (Schultz), 128*fig.*
authority obedience, 37–39, 87, 88
automobiles, 24, 57, 61, 102, 105, 157. *See also* transportation
autonomous self, 40–44. *See also* individualism; individualistic worldviews; self-absorption; self-enhancement

Bangladesh, 7
Baron, Robert, 35
barriers, 72–75; actions and, 67, 72; affects/cognitions/motivations and, 73fig.; cultural lenses and, 106; ego orientations and, 72–73; emergencies/event interpretations (stage 2) and, 82*fig.*; event notice (stage 1) and, 75*fig.*; failure of actions and, 67; habituation and, 80; helping and, 25, 72–73, 78, 104–6; ignorance and, 80–81; land ethic worldview and, 131, 147; mastery-oriented individualistic worldviews and, 72, 78–79, 106; mitigation and, 11, 72–73; natural world/nature connectedness and, 72, 79; proenvironmental behaviors and, 27, 72–75; psychology/as foundational science and, 28, 67; self-absorption/self-enhancement and, 72–73, 78; separateness and, 72; United States worldview and, 73*fig.*. *See also* obstacles; *individual barriers*
behavior changes, xvii, 24, 101, 134, 135, 149–50. *See also* human behaviors; lifestyles; proenvironmental behaviors
beliefs, xx, 15, 17–18, 40, 43, 89, 90–91, 93*fig.*, 96. *See also* just-world beliefs; *individual worldviews*
belonging, 44, 55, 60–61, 66, 118, 138–39, 140, 144, 162
"Beyond Individualism/Collectivism: New Cultural Dimensions of Values" (Schwartz), 47*fig.*, 112*fig.*
biofuels, 156, 160–61
biophilia, 116, 122–23, 153, 172
biospheric orientations, 42–43, 110, 112, 126–27, 128*fig.*, 138, 145*fig.*
birds, 77–78
Biswas-Diener, R., 139
Boer, J. de, 122–23, 129–30, 134, 137–38, 144
Boersema, J. J., 122–23, 129–30, 134, 137–38, 144
Boschetti, F., 127
Boston, 7
boycotts, 133*fig.*, 134–35
Brown, Jerry, 6
Brundtland Report (WCED 1987), 26
Bruner, Jerome, 18
Brunswik, E., 12–13, 15, 19
Buro, K., 141, 146
Burroughs, J. E., 60

California, 5, 6, 8, 15–16, 74, 165, 167
Campbell, Keith, 93, 99
Canada, 8
Capaldi, C. A., 142
capitalism, 22–23, 51*fig.*, 86–87. *See also* materialism; United States worldview; wealth

carbon (CO2) emissions: actions and, 8, 155; cause and effect magnitude and, 96–97; climate change and, 24, 155; consumerism and, 55, 57; cosmetic v curative and, 8–9, 24; environmental dashboards and, 155; greenhouse gases, 3, 9; messages and, 105; natural world and, 57, 172–73; the Oberlin Project and, 153, 157; Paris climate agreement and, 24; technology and, 24, 57; temporal/spatial proximities and, 95–96; United States and, 57; wealth and, 55
carbon taxes, 9, 24, 127, 132
care: distance and, 42; egoism and, 55; humans and, 173; individualism and, 44; Land Ethic Scale (Mayer and Frantz) and, 126–27, 136; land ethic worldview and, 145–46; natural world and, 112, 126–27, 136, 145–46, 171, 173; nature connectedness and, 34–35, 171; Schwartz, Shalom and, 53, 112, 171. *See also* love; Love and Care for Nature Scale
causal attributions, 93*fig.*, 94, 95, 96
cause and effect magnitude, 93*fig.*, 96–97, 104
Central Asia, 6
Central Park, 167
certainties, xxiii, 20, 30. *See also* uncertainties
Certificate of Sustainable Agriculture Program (Lorain Community College), 154
Chai, Hui Yi, 92
changes, xxiv, 106–7, 108–9, 113, 149–50, 152. *See also* alternative worldviews; behavior changes
Chasing Ice (Orlowski, 2012), 89, 105
Chen, Z., 140
Chenoweth, E., 172
Chicago, 168
children, 43, 147–48, 154, 155, 158, 159, 165–67, 173
Children and Nature Network, the, 165–67
China, 6
Chow, J. T., 140
Christianity/churches, 168–70. *See also* religion; spirituality
City Fresh, 154–55
civil rights movement, 172
clarity: actions and, 151; alternative worldviews and, 109; ambiguities and, 30–31; cultural lenses and, xx, xxii, 106; interconnectedness and, 172–73; interest groups and, 104; just-world beliefs and, 64; mastery-oriented individualistic worldviews and, 106; natural world/nature connectedness and, 113, 172–73; schemas/worldviews and, 29; science and, 109; the Seekers (cult group) and, 31–32; social comparisons and, 88; systems thinking and, 65–66; United States worldview and, 106; unjustified clarity distortions, 19, 29–32. *See also* perception v reality
Clayton, Susan, 10, 118–21
Cleveland, 154, 167–68
cleverness v wisdom, 100
climate change, 67–107; actions and, 3; analytic thinking and, 62, 152; ancient brains and, 76–77; behavior changes and, 149–50; capitalism and, 87; carbon (CO2) emissions and, 24, 155; causal attributions and, 96; cause and effect magnitude and, 96–97; climate chaos/destabilization and, 2–3; collective efficacy and, 99; communities and, 142, 151, 159; conservatives and, 105–6; consumerism and, 34*fig.*; cosmetic v curative and, 8–9; creativity and, 99–100; cultural lenses/scripts/worldviews and, xx, 24, 39–40; deaths and, 5; denial and, 83, 91–92, 98; depleted directed attention and, 99; dis/connection and, 72; disinformation campaigns and, 88; distance and, 37, 72, 92; distortions and, 1, 25, 64–65; egalitarianism and, 127; emergencies and, 88–92;

climate change *(continued)*
environmental movements/ grassroots movements and, 24, 151; environmental threats and, 9, 67, 76, 83, 152; escapism and, 80; failure of actions and, 25–26, 72; fatalism and, 98; fears and, 91; global warming, 2–3, 105, 170; habituation and, 80; health and, 8; helping and, 25–26; holistic/systems thinking and, 62, 100; humans/human behaviors/ systems and, 9–10, 11, 27, 34*fig.*, 73, 145*fig.*, 146; ignorance and, 98; indifference and, 34; individualistic worldviews and, xxiii, 152; information and, 89; interest groups and, 87–88; just-world beliefs and, 64, 96; land ethic worldview and, 149; land use and, 34*fig.;* Latane and Darley's (1970) model and, 73; legislation/ political activism/politics and, xix, 172; liberals and, 86–87, 105; mastery-oriented individualistic worldviews and, 72, 106, 152; materialism and, 33–34; Mayer, F. Stephan and, 100; messages/ mistrust and, 85, 105–6; moral distance and, 64; moral domains and, 87; nations and, 96, 105–6; natural world and, 72; the Oberlin Project and, 152, 158; obstacles and, 66; optimism bias and, 83; place attachments and, 79; pluralistic ignorance and, 90–91; policies and, xix; proenvironmental behaviors and, 26, 67; progress and, 100; psychological distance and, 92; psychology and, xix, xxiii, 9–10, 25, 34*fig.*, 66, 67; religion and, 168–70; relocations and, 7–8; responsibility diffusions and, 94–95; right-wing political ideologies and, 85–87, 103, 105–6; science and, 2, 88, 89; self-efficacy and, 98–99, 142; sensory limitations and, 75–76; sight and, 1; social norms and, 90–91; storms and, 96–97; superiority and, 53; technology and, xv–xvi, 9; temporal/spatial proximities and, 95–96; tipping points and, 4, 149; tragedies and, 2–9, 29; Transition Town Totnes and, 160–61; United States and, 8, 90; United States worldview and, xx, xxiii, 12, 21, 25, 27, 29, 33, 64–66; urban/lifestyles and, 33, 99; weather events and, 3; worldviews and, xix, 12, 27. *See also* psychology as a foundational science; "Psychology's Contributions to Understanding and Addressing Global Climate Change" (Swim et al.)
climate chaos/destabilization, 2–3. *See also* societal collapse
CO2 levels, 75–76, 79, 83–84, 85, 100. *See also* carbon (CO2) emissions
coastal cities, 7
cognitions, 10*fig.*, 34*fig.*, 40, 43, 73*fig.*, 145*fig.*
cognitive dissonance theory, 89–90
Cohn, Steve, 22–23, 58, 60
Collapse (Diamond), 20–21, 147–48
collective actions, 24–25, 97, 105
collective efficacy, 97*fig.*, 98–99, 108
collective identities, 119
collective vision, xx–xxiii. *See also* sight
Collectivism (Schwartz model), 47*fig.*, 48, 112*fig.*
Colorado, 5, 77
commercialism, xvi. *See also* consumerism; materialism
commitment, 97*fig.*, 116, 120, 134, 135
Commitment to the Environment Scale, 118–19, 120, 124*fig.*, 126*fig.*, 128*fig.*, 137*fig.*, 140*fig.*
commonsense psychology, 18
communities: actions and, 168; behavior changes and, xvii; the Children and Nature Network and, 166; climate change and, 142, 151, 159; collective efficacy and, 99; environmental threats and, 99; Great Transitions and, 108, 150; green spaces and, 168; individualistic worldviews and, 41*fig.*, 45*fig.;* interconnectedness

and, 106; land ethic worldview and, 54*fig.*, 110–11, 142, 149, 163–64; natural world and, 110–11, 114–15, 116, 142, 152; religion and, 168–69; resiliencies and, 4, 159, 172; Schumacher College and, 162–63; social risks and, 102; *Vitamin N: The Essential Guide to a Nature Rich Life* (Louv) and, 166; wealth and, 59; well-being and, 106, 138–39, 140, 160–61. *See also* Oberlin Project, the; Transition Town Totnes
compassion, 34–35, 42, 55, 64–65, 68, 109, 111, 145*fig.*
Conceptual Framework of Psychology, The, 13*fig.*
conflicts, xxii–xxiii, 6–7, 10*fig.*, 26, 28, 34*fig.*, 145*fig.*, 147–48. *See also* intergroup conflicts
Connectedness to Nature Scale. *See* Land Ethic Scale (Mayer and Frantz)
Connectedness to Nature Scale (Hedlund-de Witt et al.), 122–23, 124, 129–30, 137–38, 144
Connectedness With Nature Scale (Dutcher et al.), 122–23, 124*fig.*, 129, 131, 134, 136, 137
connections. *See* dis/connection; nature connectedness
conservation groups, 133*fig.*, 136. *See also* environmental organizations
conservatives, 87, 88, 89, 92, 105–6. *See also* right-wing political ideologies
constructive processes of the mind, 12–16, 19, 41*fig.*, 45*fig.*, 54*fig.*, 111*fig.*
consumerism, xvi, 10*fig.*, 34*fig.*, 55–57, 73*fig.*, 106, 129–30, 145*fig.*
consumer practices, 8–9, 20–21, 113, 132–36, 138. *See also* materialism
cooperation, xvi, 108, 141, 145*fig.*, 146, 170, 172
Copernicus, 22
cosmetic v curative, 8–9, 24
cost/benefit analyses, 68–69, 101–4, 105
Crabtree, Tim, 162
creativity, 59–61, 66, 97*fig.*, 99–100, 108, 143–44, 148, 159
crops, 4, 5–6, 154. *See also* agriculture; farms; organic food
cultural distance, 36, 50
cultural lenses: actions and, xx; autonomous self and, 40; barriers and, 106; clarity and, xx, xxii, 106; climate change and, xx, 39–40; conflicts and, xxii–xxiii; cultural knowledge/scripts and, xx, 39; dis/connection and, 39; distortions and, xx, xxii, 29–30, 32; environmental threats and, xx; errors/misjudgments and, xxi, xxiii; Great Transitions and, 148; harmony and, xx, xxiii, 33; human/nature split and, 45; land ethic worldview and, 118, 138, 142, 146; mastery-oriented individualistic worldviews and, 45; Muller-Lyer illusion and, xxii; natural world and, xx, xxiii, 39–40; perception v reality and, 19; schemas and, 30; separateness and, 106; United States and, 40; worldviews and, xxii–xxiii, 30, 32, 106
cultural schemas, 13*fig.*, 17–21, 30
cultural scripts, 37–40, 161
cultural worldviews: actions and, xx, 39; climate change and, 24; *Collapse* (Diamond) and, 21; cultural knowledge/cultural scripts and, xx, 39; distortions and, 19–20, 32; Francis (pope) and, 169; human systems and, 10*fig.*; just-world beliefs and, 63; knowledge and, 12–13; natural world and, 169; psychology/as foundational science and, 11, 17–19, 22; "Psychology's Contributions to Understanding and Addressing Global Climate Change" (Swim et al.) and, 10*fig.*; schemas and, 30; Schwartz, Shalom and, 45–53; wealth and, 65. *See also* United States worldview
cultures: actions and/definitions of, 17–18; distortions/Easter Island and, 20; folk psychology and, 18; Fromm, E. and, 22; happiness and, 17, 20,

cultures *(continued)*
58, 161; human behaviors and, 18–19; individualism and, 97, 99; love/marriage and, 44; natural world and, 53, 138; psychologists and, 17–19, 20; Schwartz, Shalom and, 46, 51–52; technological devices and, 78; Transition Town Totnes and, 161; wealth and, 51*fig.*, 58; well-being and, 17, 19; worldviews and, 32, 43

Darley, J. M., 67–74, 94–95, 101, 130, 131
Davis, J. L., 118–19
deaths, 4–5
deforestations, 8, 20
Deiner, E., 139
denial, 26, 62–64, 82*fig.*, 83, 85–86, 91–92, 98
depleted directed attention, 97*fig.*, 99
Diamond, Jared, 20–21, 147–48
diffusion of responsibility, 70, 93*fig.*, 94–95. *See also* personal responsibilities (stage 3)
Dion, K. K. and K. L., 44, 55
dis/connection: actions and, 68–69, 143; belonging and, 66; clarity and, 172–73; climate change and, 72; compassion and, 34–35, 42, 68; Connectedness to Nature Scale and, 114–15; cultural lenses and, 39; distance and, 35; economics and, 66; emergencies and, 68–69, 73–74; Great Transitions and, 108; harm and, 34–35; helping and, 72; holistic thinking and, 113; humans and, 35, 40; land ethic worldview and, 151; *Last Child in the Woods: Saving Our Children from Nature-Deficit Disorder* (Louv) and, 165–66; Latane and Darley's (1970) model and, 68–69; mastery-oriented individualistic worldviews and, 45; natural world and, 33, 34*fig.*, 37, 39, 40, 45, 55, 61, 65–66, 79, 94, 153; place attachments and, 79; politics and, 66, 170; psychology/psychological impunity and, 35, 66; technological devices and, 94. *See also* distance; indifference; nature connectedness; separateness
diseases, 8, 10*fig.*, 34*fig.*, 44, 73*fig.*, 145*fig.*
disharmony, xix, 33. *See also* harmony
disinformation campaigns, 82*fig.*, 87–88
distance, 44–52; aggression and, 35; authority obedience and, 39; autonomous self and, 40–44; climate change and, 37, 72, 92; cultural distance, 36, 50; dis/connection and, 35; emergencies and, 92; ethnicities and, 36; harm and, 36–37, 57; human/nature split and, 37, 42; humans and, 35, 169; hypothetical/social/spatial/temporal distance, 92; individualism/individualistic worldviews and, 40–44, 50; *On Killing: The Psychological Cost of Learning to Kill in War and Society* (Grossman) and, 36; land ethic worldview and, 138; mastery-oriented individualistic worldviews and, 45, 121; mechanical distance, 36, 42; moral distance, 36, 52, 62–64; natural world and, 34–37, 44, 52, 57, 61, 72, 121, 123, 127–29, 169; object self and, 42; psychological distance, 82*fig.*, 92, 152; race and, 36; scales and, 121, 127–29; social distance, 36, 92; superiority and, 55; United States and, 49, 170; United States worldview and, 34–37, 40–52, 127; urban lifestyles and, 36, 52
distortions, 17–19; actions and, 1, 20; climate change and, 1, 25, 64–65; collective vision and, xxi–xxii; cultural lenses/schemas/worldviews and, xx, xxii, 16–17, 19–20, 29–30, 32; Easter Island and, 20; individualistic worldviews and, 44; just-world beliefs and, 63; mastery-oriented individualistic worldviews and, 142–43, 151; natural world and, 64–65; old woman/young woman illusion (Hill) and, 30; psychology as

a foundational science and, 17–19; the Seekers (cult group) and, 31–32; tragedies and, 29; United States worldview and, xx, 12, 21, 25, 29, 33, 64–65, 66, 106; unjustified clarity distortions, 17–20, 29–32; worldviews and, 16–17, 19–20, 29, 32, 64–65, 106
Doherty, Thomas, 10
Donne, John, 41
Dopko, R. L., 141, 142, 146
droughts, 3, 4, 5–6, 10*fig.*, 34*fig.*, 73*fig.*, 77, 97, 145*fig.*, 159
Dutcher, D. D., 122–23, 129, 131, 134, 136, 137

Eaarth, 3–5, 11, 26, 28, 141, 146, 169
Easter Island, 20
Ebbinghaus illusion, xxi, xxiii, xxiv
ecological behaviors, 101, 136, 137*fig. See also* proenvironmental behaviors
ecological designs, xvii. *See also* Oberlin Project, the; Transition Town Totnes
ecological footprints, 60. *See also* carbon (CO2) emissions
ecological worldviews, 54*fig.*, 133*fig.*, 138, 139, 145–46. *See also* land ethic worldview
economics, 7, 24, 66, 87, 103, 105, 153–56, 157–58, 162
economic systems, xix, 23
eco orientations, 52–55
ecosystems, 5, 8, 43, 62
education, xvii, 81, 139, 147–48, 153, 162–63. *See also* Oberlin Project, the; Transition Town Totnes
Edwards, Andree, 163–64
Egalitarian Commitment (Schwartz model), 46–48, 50, 53, 54*fig.*, 112*fig.*
egalitarianism: carbon taxes/climate change/energy-efficient technologies/environmentalism and, 127; Great Transitions and, 108; Land Ethic Scale (Mayer and Frantz) and, 115, 120, 127; land ethic worldview and, 52*fig.*, 110–12, 126–27, 130, 152, 157, 170, 172; natural world and, 122; the Oberlin Project and, 154, 157; policies and, 127; Schumacher College and, 162–63; self-transcendence and, 171; spirituality and, 144; Transition Town Totnes and, 160–62; transportation and, 127
egoism, 48, 49, 53, 60–61, 80, 146, 162
ego orientations, 52–55; barriers and, 72–73; hierarchy and, 54; human/nature split and, 54; humans and, 55; ignorance and, 80; land ethic and, 112; Land Ethic Scale (Mayer and Frantz) and, 126–27; land ethic worldview and, 126–27, 138, 146; mastery-oriented individualistic worldviews and, 45; natural world and, 53; psychologists and, 55; self-transcendence and, 55; superiority and, 49, 53–55; United States and, 49; United States worldview and, 34*fig.*, 52–55, 65, 72–73, 138
Einstein, A., 65
electricity, 24, 57, 133*fig.*, 135, 154, 156, 159
Ellison, Ralph, 32
Embeddedness (Schwartz model), 46–47, 112*fig.*
emergencies, 81–92; barriers and, 82*fig.*; climate change and, 88–92; CO2 levels and, 79; denial and, 83, 91–92; dis/connection and, 68–69, 73–74; disinformation campaigns and, 87–88; distance and, 92; event interpretations (stage 2) and, 81–92; failure of actions and, 71; helping and, 81; ignorance and, 90–91; interest groups and, 104; Latane and Darley's (1970) model and, 68–70; mistrust and, 84–85; nature connectedness and, 73–74; optimism bias and, 82–83; pleasant weather incongruity and, 81–82; proenvironmental behaviors and, 78, 81–92; right-wing political ideologies and, 85–88; selective exposures and, 89–90; self-absorption and, 78;

emergencies *(continued)*
social comparisons and, 71, 88; social norms and, 90–91; suprahuman powers and, 84; technosalvation and, 83–84; uncertainties and, 81, 88
Emotional Affinity toward Nature Scale, 116–17, 120, 124*fig.*, 126*fig.*, 128*fig.*, 137*fig.*, 140*fig.*
energy-efficient technologies, 24, 86, 127, 156, 157, 159, 160, 170. *See also* green technologies
energy use, 73*fig.*, 155. *See also* carbon (CO2) emissions; carbon taxes; electricity
England. *See* Great Britain; Schumacher College; Transition Town Totnes
environmental causes, 136–37. *See also* environmental organizations; proenvironmental behaviors
environmental dashboards, 155–56, 158
Environmental Identity Scale, 118–21, 124, 126*fig.*, 128*fig.*, 133–34, 137*fig.*, 140*fig.*
environmentalism, 86, 127
environmentalists, 20, 84, 85–87, 88, 102, 104, 105, 119*fig.*, 136, 168. *See also* individuals
Environmental Justice Foundation, 7–8
environmentally friendly products, 133*fig.*, 134–35
Environmental Movement Activism Subscale, 136–37
environmental movements, 24–25, 136–37, 166, 168, 169. *See also* proenvironmental behaviors; *individual movements*
environmental organizations, 92, 132–33, 136–38, 150, 169–70
environmental perspectives, 131
environmental psychology, 23, 44
environmental rights, 53. *See also* rights of nature
environmental systems: economic systems and, xix; harmony and, xix, xxiii, 11, 22, 26, 33, 66, 72, 109, 110, 138, 143, 145–46, 173; humans and, 43; human systems and, xix, xxiii, 10–11, 26, 33, 34*fig.*, 66, 72, 73*fig.*, 145–46, 173; land ethic worldview and, 110; political systems and, xix; "Psychology's Contributions to Understanding and Addressing Global Climate Change" (Swim et al.) and, 10*fig.*, 145*fig.*; United States worldview and, xix
environmental threats: ancient brains and, 76–77; climate change and, 9, 67, 76, 83, 152; collective efficacy and, 98–99; communities and, 99; creativity and, 99–100; cultural lenses and, xx; denial and, 91–92; depleted directed attention and, 99; just-world beliefs and, 96; mastery-oriented individualistic worldviews and, 106; mistrust and, 84; nature connectedness and, 130–31; the Oberlin Project and, 152, 156; optimism bias and, 83; overwhelmed feelings and, 98; psychology and, 104; Transition Town Totnes and, 161; United States worldview and, 33; urban lifestyles and, 99
environment protections, 46–48, 50, 51*fig.*, 60, 79, 133*fig.*, 135
environments, 20, 42–43, 61, 87, 130–31, 137–38, 171. *See also* urban environments
equality, 47*fig.*, 48, 50, 52*fig.*, 53. *See also* Egalitarian Commitment (Schwartz model); egalitarianism; Harmony (Schwartz model)
errors, xx–xxi, xxiii
escapism, 75*fig.*, 80
ethnocentric beliefs, 93*fig.*, 96
Europe, 3, 4–5, 100. *See also* Schumacher College; Transition Town Totnes
event interpretations (stage 2), 68–70, 81–92; anti-intellectualism and, 85; barriers and, 82*fig.*; causal attributions and, 94; denial and, 82*fig.*, 91–92; disinformation campaigns and, 82*fig.*, 87–88; emergencies and, 81–92; failure of actions and, 71; helping and, 81; hot tub example

and, 74; ignorance and, 90–91; mistrust and, 82*fig.*, 84–85; nature connectedness and, 73–74; optimism bias and, 82–83; pleasant weather incongruity and, 81–82; psychological distance and, 82*fig.*, 92; right-wing political ideologies and, 82*fig.*, 85–88; selective exposures and, 82*fig.*, 89–90; social comparisons and, 71, 82*fig.*, 88; social norms and, 90–91; technosalvation and, 82*fig.*, 83–84; uncertainties and, 81, 82*fig.*, 88
event notice (stage 1), 68–69, 73–81, 94

failure of actions, 25–26, 67, 71, 72–74. *See also* barriers; Latane and Darley's (1970) model; obstacles; proenvironmental behaviors
fairness, 52, 87, 92. *See also* just-world beliefs
farms, xvii, 5–6, 20, 153–55, 158
fatalism, 97*fig.*, 98
fears, xvi, 91, 163
Festinger, L., 71, 88, 89–90
financial investments, 102*fig.*, 103
Finley, J. C., 122–23, 129, 131, 134, 136, 137
Fischer, R., 140, 143
floods, 7, 8, 9, 31–32. *See also* sea-level rises
folk psychology, 18
food security, 5, 6–7, 20, 159, 162. *See also* local food economies
forests, xvii, 5, 8, 20
fossil fuels, 3, 88, 104, 153, 154, 160–61
France, 5. *See also* Europe
Francis (pope), 169
Francis, A. J. P., 140, 144
Francis, Saint, 169
Frantz, Cindy, 114–15, 125, 126–27, 128–29, 130, 131, 132–33, 135, 143, 156. *See also* Land Ethic Scale (Mayer and Frantz)
free markets, 86–87. *See also* capitalism; consumerism
Freud, Sigmund, xvi, 22, 35
Fromm, E., 22, 116
fuel-efficient cars, 24, 102, 105, 157. *See also* transportation
functional risks, 102
"Funds, Friends, and Faith of Happy People, The" (Myers), 56, 59*fig.*

Gardner, Gary, 168
Geiger, N., 90–91
Genovese, Kitty, 67, 68–70, 75
Gentile, Brittany, 93, 99
George Jones Farm, 154–55, 158. *See also* Oberlin Project, the
Germany, 36, 37–38, 83. *See also* Europe
Gifford, Robert, 9, 10, 84
global warming, 2–3, 105, 170. *See also* climate change; temperature changes
goals, 4, 28, 58–59, 102*fig.*, 103–4, 151, 152, 154, 171–72
Goodwin, Brian, 161
Gould, S. J., 54
government interventions, 87, 133*fig.*, 137–38, 157, 172. *See also* politics
Grapes of Wrath, The (Steinbeck), 170–71
grassroots movements, 151–52, 166. *See also individual projects*
Great Britain, 3, 83. *See also* Europe; Schumacher College; Transition Town Totnes
Great Transitions, 106–7, 108–9, 148–50. *See also* land ethic worldview
Green, J. D., 118–19
Green Abundance (blog), 163
green buildings, 153. *See also* smarter homes
Greenfaith, 169–70
greenhouse gases, 3, 9. *See also* carbon (CO_2) emissions
Green Scare, 85–86, 102
green spaces, 157, 158–59, 167–68, 172
green technologies, 102, 105, 153. *See also* energy-efficient technologies; smarter homes
Grossman, Dave, 36
groups, 26, 41–42, 51*fig.*, 95, 96, 99, 102, 146–47. *See also* collective actions; collective efficacy; human systems

habitability, 5–6, 7
habitats, 20
habituation, 75*fig.*, 80, 100–101, 149
happiness: cultures and, 17, 20, 58, 161; "The Funds, Friends, and Faith of Happy People" (Myers), 56; income and, 56, 58–59; individualistic worldviews and, 44; interconnectedness and, 106; Land Ethic Scale (Mayer and Frantz) and, 139–41; land ethic worldview and, 26, 140, 142, 146; mastery-oriented individualistic worldviews and, 106, 142–43; materialism and, 56, 58–61, 142–43; nations and, 58; natural world/ nature connectedness and, 133–34, 142; scales and, 139–41; Transition Town Totnes and, 161; United States and, 58–59; United States worldview and, 21, 22, 25, 33, 55–61; wealth and, 58–59, 65. *See also* well-being
Harding, Stephan, 161
harm: conservatives and, 105; cultural scripts and, 37; dis/connection and, 34–35; distance and, 36–37, 57; humans and, 35, 50, 151; interconnectedness and, 171; land ethic worldview and, 138, 151; liberals and, 87, 92, 105; natural world and, 22, 34–35, 36–37, 42, 50, 52, 57–58, 93–94, 124, 127, 129–30, 151, 169; scales and, 124
harmony: alternative worldviews and, 22; cultural lenses/worldviews and, xx, xxiii, 28, 33, 65–66; disharmony, xix, 33; environmental/human systems and, xix, xxiii, 11, 22, 26, 28, 33, 66, 72, 108–9, 110, 138, 143, 145–46, 173; humans and, 108–9, 146–47, 162, 172; interconnectedness and, 147; Land Ethic Scale (Mayer and Frantz) and, 146; land ethic worldview and, 110, 126–27, 130, 138, 143, 147; Louv, Richard and, 172; natural world and, 28, 51*fig.*, 65, 108, 162, 171, 172; spirituality and, 144
Harmony (Schwartz model), 46–48, 50–51, 53, 112*fig.*
H-bomb, 100
health, 8, 44, 105, 140, 146, 165–66. *See also* diseases
heat waves, 3, 4–5. *See also* droughts
Hedlund-de Witt, A., 122–23, 129–30, 134, 137–38, 144
Heiman, R., 18–19
Heine, S. J., 17
helping, 67–74; barriers and, 25, 72–73, 78, 104–6; causal attributions and, 94; climate change and, 25–26; cost/ benefit analyses and, 68; dis/ connection and, 72; emergencies/ event interpretations (stage 2) and, 81; Genovese, Kitty and, 67, 68–70, 75; interconnectedness and, 171; land ethic worldview and, 146; Latane and Darley's (1970) model and, 67–74, 94–95, 101, 130, 131; natural world and, 34, 130; obstacles and, 104–7; psychology and, 25–26, 34, 67, 72; responsibility diffusions and, 94–95; self-absorption and, 78; social psychology and, 171; "we" sense and, 130; worldviews and, 72. *See also* actions
helping obstacles. *See* obstacles
hierarchy, 40, 44–52, 53–54, 108, 112, 115*fig.*
Hierarchy (Schwartz model), 46–51, 53, 112
Hill, W. E., 14–15
Himalayas, 6
Hitler, Adolf, 36
Hofstadter, Richard, 85
Hofstede, Gert, 42
holistic thinking, 40, 61–62, 100, 112–13, 143–44, 147, 152–53, 156, 160, 169–70. *See also* systems thinking
Holocene period, 4, 149
Hopkins, Rob, 160–61
hot tub example, 74–75
Howard, George, 10
Howell, A. J., 141, 146
human behaviors: analytic thinking and, 61; ancient brains and, 76; beliefs

and, 89; climate change and, 9–10, 27, 34*fig.*; cultures and, 18–19; habit/commitment and, 100–101; individualistic worldviews and, 43; Lewin, Kurt and, 43; materialism and, 60; misperceptions and, 43; natural world and, 55, 60; social norms and, 90; systems thinking and, 62; temporal/spatial proximities and, 95. *See also* habituation; proenvironmental behaviors
human/nature split: Christianity and the, 169; compassion and the, 64–65; cultural lenses and the, 45; distance and the, 37, 42; ego orientations and the, 54; Environmental Identity Scale and the, 120; land ethic worldview and the, 147; Leopold, Aldo and the, 109; mastery-oriented individualistic worldviews and the, 45, 52, 93–94; materialism and the, 57–58; object self and the, 42; Roszak, T. and the, 130; superiority and the, 50; United States worldview and the, 64–65. *See also* natural world
human rights, 172
humans: aggression and, 35; Anthropocene period and, 4; biophilia and, 116; care/love/respect and, 173; climate change and, 34*fig.*, 73, 145*fig.*, 146; cultural schemas and, 17; dis/connection and, 35, 40; distance and, 35, 169; distortions and, 16–17; ego orientations and, 55; environments/environmental systems and, 42–43; escapism and, 80; folk psychology and, 18; Francis (pope) and, 169; *The Grapes of Wrath* (Steinbeck) and, 171; greenhouse gases and, 3; harm and, 35, 50, 151; harmony and, 108–9, 146–47, 162, 172; hierarchy and, 54, 112; interconnectedness and, 110, 171–72; land ethic worldview and, 110–11, 145–47, 170, 171–72; misperceptions and, 17; natural world and, xv–xvii, 44, 45, 50, 52–55, 100, 104, 108, 110–11, 116, 123*fig.*, 126, 151–52, 162, 171–72; nature connectedness and, 116, 161, 166, 171–73; the Oberlin Project and, 157; place attachments and, 79; responsibility diffusions and, 94–95; self-transcendence and, 55; separateness and, 44, 45; social distance and, 36; technology and, xv–xvi. *See also* happiness; well-being
human systems: climate change and, 9–10, 11, 34*fig.*; cultural worldviews and, 10*fig.*; environmental systems and, xix, xxiii, 10–11, 26, 33, 34*fig.*, 66, 72, 73*fig.*, 145–46, 173; Great Transitions and, 108–9; greenhouse gases and, 9; harmony and, xix, xxiii, 11, 22, 26, 28, 33, 66, 72, 108–9, 110, 138, 143, 145–46, 173; land ethic worldview and, 110, 145–47; mastery-oriented individualistic worldviews and, 145–46; natural world and, 28. *See also* humans
hurricanes, 3, 97. *See also* storms

idea formations (stage 4), 68–69, 74–75, 97–101
ignorance, xv, 75*fig.*, 80–81, 90–91, 97*fig.*, 98
illnesses, 44, 165–66. *See also* diseases; health
impropriety denials, 62–64
Inclusion of Nature in Self Scale, 127–29, 141
income, 56, 58–59. *See also* wealth
India, 6
indifference, 33–37; climate change and, 34; dis/connection and, 72–73; land ethic worldview and, 138, 151; mastery-oriented individualistic worldviews and, 123; natural world and, 33–37, 39, 42, 50, 52, 72–73, 93–94, 122, 123; psychological impunity and, 72; self-absorption and, 79; tragedies and, 34; United States worldview and, 34–37

individualism: analytic thinking and, 61; belonging/separateness and, 44; care/love/marriage and, 44, 55; collective actions and, 97; cultures and, 97, 99; distance and, 50; materialism and, 56, 61; natural world and, 50, 57, 151; religion and, 168; Schwartz, Shalom and, 47, 49, 112*fig.*; Transition Town Totnes and, 161. *See also* "I" sense; self-absorption; self-enhancement
Individualism (Schwartz model), 47*fig.*, 48, 112
Individualism and Collectivism: Theory, Method, and Applications (Kim et al.), 112
individualistic worldviews, xxiii, 40–44, 52, 53–55, 77–78, 152, 168. *See also* mastery-oriented individualistic worldviews
industrialization, 20, 27, 42, 75–76, 77, 79–80, 94, 127
information, 17–18, 21, 30, 81, 82*fig.*, 87–88, 89–90, 92
in-group loyalties, 87, 92, 105
Inspiring Progress: Religion's Contributions to Sustainable Development (Gardner), 168
Intellectual Autonomy (Schwartz model), 46–47, 112*fig.*
intercom system study, 71
interconnectedness, 147–48; actions/cooperation and, 172; children and, 43; clarity and, 172–73; communities and, 106; goals and, 171–72; *The Grapes of Wrath* (Steinbeck) and, 171; happiness and, 106; harm/helping and, 171; harmony and, 147; humans and, 110, 171–72; land ethic worldview and, 110, 148, 170, 171–72; Leopold, Aldo and, 113; lifestyles and, 65–66; natural world and, 106, 110, 112, 116, 152, 155; perspective-taking and, 147; resiliencies and, 172; Schumacher College and, 163; Transition Town Totnes and, 161. *See also* dis/connection; nature connectedness
interdependence theory, 118
interest groups, 88, 104
Interface Between Psychology and Global Climate Change (APA Task Force), 10
Interfaith Power and Light, 170
intergroup conflicts, 10*fig.*, 21, 34*fig.*, 73*fig.*, 145*fig.*, 147–48, 160
internal lenses. *See* schemas
intraself discrepancies, 91–92
Invisible Man (Ellison), 32
"I" sense, 40–42, 44–45, 52, 54*fig.*, 79, 93–94, 99, 110–11. *See also* mastery-oriented individualistic worldviews; self-absorption; self-enhancement

Jacoby, Susan, 85
Johnson, J B., 122–23, 129, 131, 134, 136, 137
Juang, L., 17, 20
justice, 7–8, 52–53, 169. *See also* rights of nature; social justice
just-world beliefs, 40, 62–64, 93*fig.*, 96, 106

Kals, Elisabeth, 116–17, 164–65
Kamitsis, I., 140, 144
Kanner, Allen, 22–23, 58, 60
Kansas, 5
Kaplan, Steve and Rachel, 99
Kasser, Tim, 22–23, 58–59, 60
Keech, Marian, 31
Keeling Curve, 76*fig.*
Kellert, S.O., 116
King Lear (Shakespeare), 1–2, 12
kinship, 109–10, 111*fig.*, 114–15, 120, 122, 144. *See also* egalitarianism; interconnectedness; land ethic worldview
Kitayama, S., 18–19
knowledge, xx, 12–13, 15–16, 80. *See also* ignorance; information
Kohn, Alfie, 60
Krauthammer, Charles, 86
Krugman, Paul, 24
Kyrgyzstan, 6

land ethic, 110, 112, 120, 122–23
Land Ethic Scale (Mayer and Frantz), 114–18, 122–29, 131–41; consumerism

and the, 130; creativity and the, 143; distance and the, 127–29; egalitarianism and the, 115, 120, 127; ego orientations and the, 126–27; electricity and the, 135; environmental organizations and the, 136–37; environmental perspectives/perspective-taking and the, 131; happiness and the, 139–41; harmony and the, 146; holistic thinking and the, 143; land ethic worldview and the, 114–18, 120, 131, 158; lifestyles and the, 114, 125, 132–33, 135–36; natural world and the, 127, 132–33, 158, 164; political activism and the, 138; proenvironmental behaviors and the, 125, 131–33, 135–38; scales and the, 120, 122–26; self/nature divide and the, 128–29; self-transcendence/self-enhancement and the, 126; spirituality and the, 144–45; urban lifestyles/validity and the, 114; well-being and the, 139–41, 144–45

land ethic worldview, 110–73; actions and the, 132, 149, 151–73; adaptations and the, 146; alternative worldviews and the, 110; barriers and the, 131, 147; behavior changes and the, 149; the Children and Nature Network and the, 165–67; churches and the, 168–70; climate change and the, 149; communities and the, 54*fig.*, 110–11, 142, 149, 163–64; constructive processes of the mind and the, 54*fig.*; consumerism/consumer practices/materialism and the, 129–30, 132–36; cooperation and the, 146, 170, 172; creativity and the, 143–44, 148, 159; cultural lenses and the, 118, 138, 142, 146; dis/connection and the, 151; distance and the, 127–29, 138; Eaarth and the, 26; egalitarianism and the, 52*fig.*, 54*fig.*, 110–12, 126–27, 130, 152, 157, 170, 172; ego orientations and the, 126–27, 138, 146; environmental dashboards and the, 156; environmental organizations and the, 132–33, 136–37; environmental systems and the, 110; environments and the, 130–31; goals and the, 152; happiness and the, 26, 140, 142, 146; harm and the, 138, 151; harmony and the, 110, 126–27, 130, 138, 143, 147; helping and the, 146; holistic thinking and the, 143–44; human/nature split and the, 147; humans/human systems and the, 110–11, 145–47, 170, 171–72; indifference and the, 138, 151; interconnectedness and the, 110, 148, 170, 171–72; "I"/"we" sense and the, 54*fig.*, 110–11; Land Ethic Scale (Mayer and Frantz) and the, 114–18, 120, 131, 158; Leopold, Aldo and the, 23, 110–13, 116–17, 125; lifestyles and the, 132–36, 138, 149; mastery-oriented individualistic worldviews and the, 110–12, 114–15, 126–27, 145–46; mitigation and the, 146; natural world and the, 114, 118, 122, 130, 146, 152, 158, 164, 170; nature connectedness and the, 110, 114, 146, 152, 171–72; the Oberlin Project and the, 152–59, 163; perception v reality and the, 54*fig.*, 111*fig.*; perspective-taking and the, 130–31, 138, 146; politics/political activism and the, 132–33, 137–38, 172; problem-solutions and the, 143–44, 148, 159; proenvironmental behaviors and the, 23, 26, 127, 131–38, 159; "Psychology's Contributions to Understanding and Addressing Global Climate Change" (Swim et al.) and the, 145*fig.*; religion and the, 169; resiliencies and the, 146; science and the, 114–15; self-efficacy and the, 122, 146; self-enhancement and the, 118, 126–27, 173; self/nature divide and the, 128–29; self-transcendence and the, 118, 126–27, 138, 146; spirituality and the, 118, 144–46; survival and the, 26; transitions to

land ethic worldview *(continued)*
the, 26, 53, 151–52, 163–64, 167, 169–70, 172; Transition Town Totnes and the, 160–64; urban planning and the, 167–68; values and the, 126; well-being and the, 23, 26, 118, 138–43, 144 45, 159
land use, 10*fig.*, 34*fig.*, 73*fig.*, 145*fig.*, 154
Last Child in the Woods: Saving Our Children from Nature-Deficit Disorder (Louv), 165–66
Latane, B., 67–74, 94–95, 101, 130, 131
Latane and Darley's (1970) model, 67–74, 94–95, 101, 130, 131. *See also individual stages*
legislation, 103, 132, 172. *See also* government interventions; politics
lens model (Brunswik), 12–13, 15, 19
Leong, L. Y. C., 140, 143
Leopold, Aldo, 23, 108–13, 116–17, 125, 130–31
Lerner, Melvin, 63
Lewin, Kurt, 43, 151
Lezak, S. B., 61–62
liberals, 86–87, 89, 92, 102, 105. *See also* right-wing political ideologies
lifestyles, 132–36; climate change and, 33; interconnectedness and, 65–66; Land Ethic Scale (Mayer and Frantz) and, 114, 125, 132–33, 135–36; land ethic worldview and, 132–36, 138, 149; natural world and, 28, 132–33, 134, 161, 172; nature connectedness and, 134; religion and, 169; status quo and, 103; sustainability and, 162; United States worldview and, 33. *See also* urban lifestyles
Limbaugh, Rush, 86, 88
local food economies, 153–55, 159, 160, 168
Lorain Community College, 154
Los Angeles, 15–16, 74, 165
Louv, Richard, 165–67, 168, 172
love, 44, 55, 116, 138–39, 164–65
Love and Care for Nature Scale, 116–17, 121, 124, 134–37
Luloff, A. E., 122–23, 129, 131, 134, 136, 137
Maldives, 7–8
Markham, Edwin, 30–31
Markus, H. R., 18–19
marriage, 44, 55, 139. *See also* love
Marshall, S.L.A., 35
Maslow, Abraham, 55
Mastery (Schwartz model), 46–51, 53, 112
mastery-oriented individualistic worldviews: analytic thinking and, 106, 152; barriers and, 72, 78–79, 106; clarity and, 106; climate change and, 72, 106, 152; consumerism and, 106; dis/connection/distance and, 45, 121; distortions and, 142–43, 151; environmental threats and, 106; Francis (pope) and, 169; goals/values/aspirations and, 103–4; happiness and, 106, 142–43; human/nature split and, 45, 52, 93–94; human systems and, 145–46; indifference and, 123; just-world beliefs and, 106; land ethic world-view and, 110–12, 114–15, 126–27, 145–46; moral distance and, 52; natural world and, 57, 79, 93, 121, 123, 129, 130, 151; the Oberlin Project and, 152; personal responsibilities (stage 3) and, 93–94; proenvironmental behaviors and, 72, 93; progress/resources and, 106; Schwartz, Shalom and, 53; self-absorption and, 78–79, 169; separateness and, 123; social comparisons and, 88; superiority and, 106; tragedies and, 94; United States and, 45, 48–50, 51*fig.*, 72–73, 93; wealth and, 106; well-being and, 118, 120, 138. *See also* individualistic worldviews
materialism, 55–61; climate change and, 33–34; commercialism, xvi; creativity and, 59–60, 99–100; disharmony and, 33; egoism and, 60–61; electricity and, 57; happiness and, 56, 58–61, 142–43; human behaviors and, 60; human/nature

split and, 57–58; individualism and, 56, 61; land ethic worldview and, 129–30; natural world and, 57, 60–61, 129–30, 138; "Psychology's Contributions to Understanding and Addressing Global Climate Change" (Swim et al.) and, 34*fig.*; self-enhancement and, 33–34, 37, 55; self-esteem/self-worth/symbolism and, 56–57, 60; superiority and, 37, 61; Transition Town Totnes and, 161; United States worldview and, 33–34, 55–61, 108, 129–30; well-being and, 58–61. *See also* consumerism; consumer practices; wealth
Matsumoto, D., 17, 20
Mauna Loa Observatory, 76*fig.*
Mayer, F. Stephan: airport example and, 97–98; beach memories and, 165; climate change and, 100; cultural worldviews and, 13–14; hot tub example and, 74–75; lifestyles and, 132–33; Los Angeles and, 15–16, 74; nature connectedness and, 77; the Oberlin Project and, 156; permanent intraself discrepancies and, 91; problem-solutions and, 143; proenvironmental behaviors and, 135. *See also* Land Ethic Scale (Mayer and Frantz)
McClure, J., 140, 143
McDonald, Rachel, 92
McKibben, Bill, 3, 24, 162
mechanical distance, 36, 42
Men against Fire (Marshall), 35
mental health, 64, 140. *See also* happiness; well-being
Mesa Verde National Park, 77
messages, 92, 105–6. *See also* mistrust
Miami, 7, 8
Milgram, Stanley, 37–39
misjudgments, xx–xxi, xxiii
Mismeasure of Man, The (Gould), 54
misperceptions: analytic thinking and, 62; CO2 levels and, 75; humans/human behaviors and, 17, 43; invulnerabilities and, 2; natural world and, 12; obstacles and, 104; Orr, David and, 24; pluralistic ignorance and, 90–91; psychology/as foundational science and, 12–15, 15–17, 22, 24; tragedies and, xxiii, 1–2; United States worldview and, 12; unjustified clarity distortions and, 19; vanities and, 1–2. *See also* perception v reality
mistrust, 82*fig.*, 84–85, 87, 91–92, 97*fig.*, 101, 105–6
mitigation: actions/alternative worldviews and, 11; barriers/obstacles and, 11, 72–73; collective efficacy and, 99; creativity and, 99–100; land ethic worldview and, 146; messages and, 105; pluralistic ignorance and, 90; psychology and, 72–73; "Psychology's Contributions to Understanding and Addressing Global Climate Change" (Swim et al.) and, 10*fig.*, 34*fig.*, 145*fig.*; right-wing political ideologies and, 85; United States worldview and, 11, 73*fig.*
Montada, Leo, 116–17, 164–65
Mooney, C., 86
moral development, 55
moral distance, 36, 52, 62–64
moral domains, 87, 105, 169–70
motivations, 10*fig.*, 11, 12, 26, 34*fig.*, 59–60, 66, 73*fig.*, 99–100, 145*fig.*
Muller-Lyer illusion, xxi–xxii, xxiii, xxiv
Murphy, S. A., 120–21, 133–34, 135, 136, 137, 143
Myers, David, 56, 59*fig.*

narcissism, 16, 52, 66. *See also* self-absorption; self-enhancement
Nasheed, Mohamed, 7–8
National Audubon Society, 166
nations, 24, 33, 50, 58, 83, 96, 100, 103, 105–6, 148. *See also individual countries*
natural resources, 51*fig.*, 121*fig.*
natural world: actions and the, 44, 81, 113, 115*fig.*, 120–21; Adam J. Lewis

natural world *(continued)*
Environmental Studies Center and the, 153; analytic thinking and the, 61; autonomous self and the, 40; barriers and the, 72, 79; belonging and the, 118, 140; carbon (CO2) emissions and the, 57, 172–73; care/love/respect and the, 112, 116–18, 126–27, 136, 145–46, 164–65, 171, 173; the Children and Nature Network and the, 165–67; children and the, 154, 155, 158, 165–67; Christianity and the, 168–70; clarity and the, 113; climate change and the, 72; *Collapse* (Diamond) and the, 20; commitment and the, 116, 120; communities and the, 110–11, 114–15, 116, 142, 152; cultural lenses/scripts/worldviews and the, xx, xxiii, 39–40, 169; cultures and the, 53, 138; dis/connection and the, 33, 34*fig.*, 37, 39, 40, 45, 55, 61, 65–66, 79, 94, 153; distance and the, 34–37, 44, 52, 57, 61, 72, 121, 123, 127–29, 169; distortions and the, 64–65; eco orientations and the, 53; egalitarianism and the, 122; ego orientations/egoism and the, 53; Einstein, A. and the, 65; environmental dashboards and the, 156; Francis (pope) and the, 169; Fromm, E. and the, 22; happiness and the, 133–34; harm and the, 22, 34–35, 36–37, 42, 50, 52, 57–58, 93–94, 124, 127, 129–30, 151, 169; harmony and the, 28, 51*fig.*, 65, 108, 162, 171, 172; helping and the, 34, 130; human behaviors/systems and the, 28, 55, 60; humans and the, xv–xvii, 45, 50, 52–55, 100, 104, 108, 110–11, 116, 123*fig.*, 126, 151–52, 162, 171–72; illnesses and the, 44, 165–66; indifference and the, 33–37, 39, 42, 50, 52, 72–73, 93–94, 122, 123; individualism and the, 50, 57, 151; individualistic worldviews and the, xxiii, 42, 53; industrialization and the, 20; interconnectedness and the, 106, 110, 112, 116, 152, 155; justice and the, 52, 53; just-world beliefs and the, 96; land ethic and the, 120; land ethic worldview and the, 114, 118, 122, 130, 146, 152, 158, 164, 170; Leopold, Aldo and the, 108–13; lifestyles and the, 28, 132–33, 134, 161, 172; mastery-oriented individualistic worldviews and the, 57, 79, 93, 121, 123, 129, 130, 151; materialism and the, 57, 60–61, 129–30, 138; mechanical distance and the, 42; misjudgments and the, xxiii; misperceptions and the, 12; moral distance and the, 52; narcissism and the, 52; the Oberlin Project and the, 152, 156; object self and the, 52; personal responsibilities (stage 3) and the, 145–46; place attachments and the, 79–80; political activism and the, 138; problem-solutions and the, 44, 143; psychological distance and the, 152; psychological impunity and the, 39–40, 52, 57–58, 72, 127, 130; psychology and the, 34; religion and the, 168–70; right-wing political ideologies and, 86–87; Schwartz, Shalom and the, 46–51, 53, 112; science and the, xxiv, 66, 109; self and the, 40, 52, 114–15, 118–19, 121, 127–29, 130; self-enhancement/self-transcendence and the, 53, 55; separateness and the, 34–37, 42, 45, 72; solace and the, 117*fig.*, 118, 119*fig.*, 120, 133–34, 149; spirituality and the, 116–17, 119*fig.*, 121, 144; superiority and the, 45, 50, 53, 61; technological devices and the, 78; United States worldview and the, 33–34, 36–37, 64–65, 127; urban lifestyles and the, 36, 42, 52, 77–78; wealth and the, 58–60, 65–66, 72, 81, 129–30; well-being and the, 44, 65, 72, 118–19, 120, 122–23, 132, 140; worldviews and the, 12, 24–25, 28, 37, 72, 106–7, 114. *See also* environment protections; human/nature

split; proenvironmental behaviors; *individual scales*
nature. *See* natural world
nature connectedness, 110–32; Anthropocene period and, 149; barriers and, 72; care and, 34–35, 171; the Children and Nature Network and, 165–67; clarity and, 113, 172–73; emergencies and, 73–74; environmental concerns/threats and, 130–31; environmental dashboards and, 156, 158; environmental organizations and, 136; event notice (stage 1)/event interpretations (stage 2) and, 73–74; *The Grapes of Wrath* (Steinbeck) and, 171; Great Transitions and, 107; green spaces and, 168; happiness and, 142; Harding, Stephan and, 161; humans and, 116, 161, 166, 171–73; individualistic worldviews and, 42; land ethic worldview and, 110, 114, 146, 152, 171–72; *Last Child in the Woods: Saving Our Children from Nature-Deficit Disorder* (Louv) and, 165–66; Latane and Darley's (1970) model and, 72–74; Leopold, Aldo and, 111, 113; lifestyles and, 134; Mayer, F. Stephan and, 77; the Oberlin Project and, 157; personal growth and, 141; political activism and, 137; proenvironmental behaviors and, 72–74, 131, 152; "Psychology's Contributions to Understanding and Addressing Global Climate Change" (Swim et al.) and, 145*fig.*; rights of nature and, 110; Schumacher College and, 162; Schwartz, Shalom and, 112, 126; spirituality and, 144, 162; Transition Town Totnes and, 160–61; urban planning and, 168; *Vitamin N: The Essential Guide to a Nature Rich Life* (Louv) and, 166; wealth and, 129–30; well-being and, 139, 140; "we" sense and, 130. *See also* dis/connection; distance; human/nature split; interconnectedness; *individual scales*
nature deficit disorder, 165–66
nature preserves, 155, 158
Nature Relatedness Scale, 120–22, 124, 126*fig.*, 128*fig.*, 133–34, 136, 137, 140*fig.*, 141, 143, 146
Nature's Due (Goodwin), 161
Nebraska, 5
Newell, Ben, 92
New Environmental Paradigm, 23, 125
New Jersey, 7
New Mexico, 5
New York, 7, 167
Nisbet, E. K., 120–21, 133–34, 135, 136, 137, 141, 143
Nixon, Richard, 85
North Carolina, 7

Oberlin Project, the, 152–60, 163–64
object self, 41–42, 52
obstacles, xxiii, xxiv, 11, 18, 66, 68, 99, 100–101, 104–7. *See also* barriers
oceans, 7, 8, 76*fig.*, 97. *See also* sea-level rises
Ogallala Aquifer, 5
Ohio, 152. *See also* Oberlin Project, the
Oklahoma, 5
old woman/young woman illusion (Hill), 14–15, 19, 29–30
Olmstead, Frederick Law, 167
On Killing: The Psychological Cost of Learning to Kill in War and Society (Grossman), 36
optimism bias, 82–83, 88
organic food, 133*fig.*, 134, 154–55, 168
Orr, David, 24, 65–66, 76, 81, 107, 153
Oskamp, Stuart, 9, 26–27
overconsumption, 8–9, 113. *See also* consumerism; consumer practices; materialism
overwhelmed feelings, 97–98, 105

Pacific Islands, 84
Paris climate agreement, 24
parks, 77, 167–68. *See also* green spaces
Passmore, H.-A., 141, 146
peace psychology, 147
perceived inequities, 102*fig.*, 103

perception v reality, 12–15, 16, 19, 45*fig.*, 54*fig.*, 94, 111*fig. See also* misperceptions; unjustified clarity distortions
Perkins, H. E., 116–17, 121, 134–37
permanent intraself discrepancies, 91
personal growth, 146. *See also* well-being
personal resiliencies, 4, 10*fig.*, 34*fig.*
personal responsibilities (stage 3), 68–71, 72, 74–75, 79, 92–97, 98, 145–46
perspective-taking, 42, 125, 130–31, 138, 146, 147
Petersen, John, 135, 156
physical risks, 101–2
place attachments, 75*fig.*, 79–80, 93*fig.*, 94, 156–57
pleasant weather incongruity, 81–82
pluralistic ignorance, 90–91. *See also* ignorance
policies, xix, 24, 85, 103, 127, 153
political activism, 59, 132–33, 136–38, 157, 172
politics: climate change and, xix, 172; dis/connection and, 66, 170; environmental movements and, 24; environments and, 137; land ethic worldview and, 132–33, 137–38; proenvironmental behaviors and, 132–33; psychological distance and, 92; psychologists and, 23; Republicans, 87; selective exposures and, 89; temporal risks and, 102. *See also* liberals; right-wing political ideologies
Ponsonby, Julia, 163
Poon, K.-T., 140
population increases, xvi–xvii, 6–7, 20–21
population mobilities, 79–80, 94
Portinga, W., 127
practical skills, 160–61
Price, J. C., 127
problem-solutions, 21, 28, 44, 62, 99, 108, 143–44, 148, 159, 169
proenvironmental behaviors, 72–101, 131–38; action abilities (stage 5)/idea formations (stage 4) and, 97–101; barriers and, 27, 72–75; climate change and, 26, 67; commitment and, 135; consumer practices/lifestyles and, 132–36; emergencies and, 78, 81–92; environmental organizations and, 136–37; environments and, 137–38; event notice (stage 1) and, 75–81; failure of actions and, 25–26, 67; land ethic worldview and, 23, 26, 127, 131–38, 159; Latane and Darley's (1970) model and, 72–74; mastery-oriented individualistic worldviews and, 72, 93; Mayer, F. Stephan and, 135; motivations and, 26; nature connectedness and, 72–74, 131, 152; obstacles and, 104–5; personal responsibilities (stage 3) and, 92–97; politics/political activism and, 132–33, 137–38; religion and, 169; scales and, 119*fig.*, 120, 122–23, 124–25, 131–38; self-absorption and, 78; self-efficacy and, 98; tokenism and, 104; United States worldview and, 26
progress, 27, 40, 51*fig.*, 55–61, 100, 106, 168
pronoun-use study, 93, 99
prosocial behaviors, 25, 67–74, 78. *See also* Latane and Darley's (1970) model
"Psychological Contributions to Achieving an Ecologically Sustainable Future for Humanity" (Oskamp), 9
psychological distance, 82*fig.*, 92, 152. *See also* distance
psychological impunity, 35, 37, 39–40, 52, 57–58, 72, 127, 130
Psychological Inquiry, 23, 51–52
psychological risks, 102
psychological well-being, 140–41. *See also* well-being
psychologists, 12, 17–19, 20, 22, 23, 55, 94. *See also* individuals
psychology: aggression and, 35; ancient brains and, xvi; barriers and, 67; behavior changes and, 149; belonging and, 66; climate change and, xix, xxiii, 9–10, 25, 34*fig.*, 66,

67; *Collapse* (Diamond) and, 21; commercialism/consumerism and, xvi; commonsense/folk psychology, 18; *The Conceptual Framework of Psychology*, 13*fig.*; cultural worldviews and, 11; dis/connection and, 66; environmental threats and, 104; failure of actions and, 25–26, 67; Genovese, Kitty and, 67; helping and, 25–26, 34, 67, 72; Lewin, Kurt and, 43; misperceptions and, 12–15, 22, 24; mitigation and, 72–73; natural world and, 34; obstacles and, 11, 66; peace psychology, 147; problem-solutions and, 62; United States worldview/worldviews and, 11. *See also* proenvironmental behaviors; social psychology
psychology as a foundational science, 9–26; barriers/conflicts and, 28; cultural schemas/worldviews and, 17–21; distortions and, 17–19; environmental movements and, 24–25; misperceptions and, 15–17, 24; perception v reality, 12–15; "Psychological Contributions to Achieving an Ecologically Sustainable Future for Humanity" (Oskamp) and, 9; "Psychology's Contributions to Understanding and Addressing Global Climate Change" (Swim et al.) and, 10–11; schemas and, 15–17; worldviews and, 21–24
"Psychology's Contributions to Understanding and Addressing Global Climate Change" (Swim et al.), 10–11, 34*fig.*, 73*fig.*, 145*fig.*, 146
Punished by Rewards: The Trouble with Gold Stars, Incentive Plans, A's, Praise, and Other Bribes (Kohn), 60

Quality of Life Scale, 140

race, 36, 54, 58–59
Rails to Trails Conservancy, 168
rebound effects, 101
recycling, 133*fig.*, 134–35, 136
Reed, A., 118–19
religion, 50, 88, 106, 168–70. *See also* spirituality; suprahuman powers
relocations, 7–8. *See also* place attachments; population mobilities
renewable energies, 154, 170
Republicans, 87. *See also* conservatives; right-wing political ideologies
Reser, Joseph, 10
resiliencies, 163–64; adaptations and, 34*fig.*, 73*fig.*, 145*fig.*; communities and, 4, 159, 172; crops and, 4; interconnectedness and, 172; land ethic worldview and, 146; the Oberlin Project and, 153, 159; personal resiliencies, 4, 10*fig.*, 34*fig.*; "Psychology's Contributions to Understanding and Addressing Global Climate Change" (Swim et al.) and, 10*fig.*; Schumacher College and, 163; *Thriving beyond Sustainability: Pathways to a Resilient Society* (Edwards), 163–64; Transition Town Totnes and, 160; *Vitamin N: The Essential Guide to a Nature Rich Life* (Louv) and, 167
resources, 51*fig.*, 106, 167. *See also* natural resources
responsibility diffusions, 70, 93*fig.*, 94–95. *See also* personal responsibilities (stage 3)
Right Kind of People, The (Markham) (poem), 30–31
rights of nature, 52–53, 110, 121*fig.*, 169
right-wing political ideologies, 82*fig.*, 85–88, 103, 105–6. *See also* conservatives; politics
Rindfleisch, A., 60
risks, 101–2
Roszak, T., 130
Russia, 5
Ryan, Richard, 22–23, 58, 60

Sand County Almanac: With Essays on Conservation from Round River, The (Leopold), 110–11
Sane Society, The (Fromm), 22

Satisfaction with Life Scale, 139
scales. *See individual scales*
schemas, 15–17, 21, 29–30. *See also* cultural lenses; cultural schemas
Schultz, P.W., 127–28, 141
Schumacher, Daniel, 116–17, 164–65
Schumacher, E. F., 162
Schumacher College, 162–63
Schwartz, Shalom, 45–53, 112, 126, 144, 171
science: actions and, xxiv, 109; alternative worldviews and, 109; clarity and, 109; climate change and, 2, 88, 89; CO2 levels and, 76, 85; disinformation campaigns and, 88; economic systems and, 23; hierarchy and, 50; land ethic worldview and, 114–15; mistrust and, 84–85; natural world and, xxiv, 66, 109; uncertainties and, 88; worldviews and, xxiv, 21–22, 23, 109. *See also* psychology as a foundational science
Scripps Institute of Oceanography, 76*fig.*
sea-level rises, 7–8, 9, 10*fig.*, 34*fig.*, 73*fig.*, 84, 145*fig.*
Seekers (cult group), 31–32
selective exposures, 82*fig.*, 89–90
self, 40–43, 52, 114–15, 118–19, 121–23, 127–30
self-absorption, 33, 53, 60–61, 66, 75*fig.*, 78–79, 169. *See also* egoism; ego orientations; individualistic worldviews
self-efficacy, 97*fig.*, 98–99, 122, 142, 144, 146
self-enhancement: barriers and, 72–73; collective actions and, 97; land ethic and, 112; Land Ethic Scale (Mayer and Frantz) and, 126; land ethic worldview and, 118, 126–27, 173; materialism and, 33–34, 37, 55; natural world and, 53; personal responsibilities (stage 3) and, 93*fig.*, 95–96; right-wing political ideologies and, 86; Schwartz, Shalom and, 47*fig.*, 48, 49, 50, 112; social justice and, 50; superiority and, 53–54, 55; temporal/spatial proximities and, 95–96; United States and, 49; United States worldview and, 33–34, 37–40, 65, 72–73; wealth and, 55; well-being and, 138
self-esteem, 17, 49, 55, 60, 102
self/nature divide, 127–29, 138. *See also* human/nature split
self-transcendence: egalitarianism and, 171; ego orientations and, 55; humans/natural world and, 55; land ethic and, 112; Land Ethic Scale (Mayer and Frantz) and, 126; land ethic worldview and, 118, 126–27, 138, 146; Maslow, Abraham and, 55; Schumacher College and, 162; Schwartz, Shalom and, 47*fig.*, 48, 112, 171; social justice and, 171; spirituality and, 144
Seligman, Martin, 8
sensory limitations, 75–76, 77, 80
separateness, 108–50; barriers and, 72; changes and, 113; cultural lenses and, 106; humans and, 44, 45; individualism and, 44; mastery-oriented individualistic worldviews and, 123; natural world and, 34–37, 42, 45, 72; object self and, 42; superiority and, 50; United States and, 49; United States worldview and, 34–37, 66, 108; worldviews and, 106. *See also* dis/connection; distance
Shakespeare, William, 1, 42
Shamin, M. R., 135
Shamin, Rumi, 156
shock study (Milgram), 37–39
Short History of Progress, A (Wright), 27
Sierra Nevada, 6
sight, xix–xxii, 1–2. *See also* clarity; collective vision
Small Is Beautiful: A Study of Economics as If People Mattered (Schumacher), 162
smarter homes, 156, 157
Smith, Edward, 2
smoke study, 70

social comparisons, 71, 82*fig.*, 88, 90
social distance, 36, 92
social justice, 47–48, 50, 52, 53, 171
social norms, 17, 90–91. *See also* cultures
social power, 47, 49, 51*fig.*
social psychology, 25, 43, 70–71, 147, 171
social risks, 102
social well-being, 141–42, 146. *See also* belonging; well-being
societal collapse, 20–21, 33, 147–48
solace, 117*fig.*, 118, 119*fig.*, 120, 133–34, 149
solar energy, 24, 153, 159, 160
South Carolina, 7
South Dakota, 5
species rights, 52–53. *See also* rights of nature
spirituality, 116–18, 119*fig.*, 120–21, 124, 144–46, 160, 162, 169–70. *See also* religion
stages. *See* Latane and Darley's (1970) model
status quo, 86–88, 103, 104
Steg, L., 127
Stegner, Wallace, xv
Steinbeck, John, 170–71
Stephan, M. J., 172
Stern, Paul, 10
Stevenson, Adlai, 85
Stites, Jessica, 163
Stoessinger, J. G., 83
storms, 34*fig.*, 73*fig.*, 96–97, 145*fig.*
Subjective Happiness Scale, 139
Summer Discovery Camp, 154, 158
superiority, 36, 37–40, 44–52, 53–55, 61, 106
suprahuman powers, 64, 83–84. *See also* technosalvation
survival, 25, 26, 33
sustainability: Brundtland Report (WCED 1987) and, 26; Certificate of Sustainable Agriculture Program (Lorain Community College), 154; *Collapse* (Diamond) and, 20–21; creativity and, 143; Eaarth and, 4; economics and, 156; education and, 162; *Inspiring Progress: Religion's Contributions to Sustainable Development* (Gardner), 168; Interfaith Power and Light and, 170; lifestyles and, 162; the Oberlin Project and, 153–54, 156; Orr, David and, 153; Oskamp, Stuart and, 26–27; "Psychological Contributions to Achieving an Ecologically Sustainable Future for Humanity" (Oskamp), 9; Schumacher College/spirituality/technology and, 162; temporal/spatial proximities and, xvii; *Thriving beyond Sustainability: Pathways to a Resilient Society* (Edwards), 163–64; water and, 20–21
Swim, Janet, 10, 34*fig.*, 73*fig.*, 90–91, 145*fig.*
system justifications, 102*fig.*, 103
systems thinking, 61–62, 65–66, 112–13

Tajikistan, 6
Tam, K.-P., 126, 128, 136–37, 138, 139
technological devices, xv, 56, 57, 75*fig.*, 78, 94, 156
technology: carbon (CO2) emissions and, 24, 57; the Children and Nature Network and, 166; climate change and, xv–xvi, 9; environmental movements and, 24; event notice (stage 1) and, 78; Francis (pope) and, 169; Goodwin, Brian on, 161; green technologies, 102, 105, 153; humans and, xv–xvi; Stegner, Wallace and, xv; sustainability and, 162. *See also* energy-efficient technologies
technosalvation, 82*fig.*, 83–84
Tegua, 7
Teixeira model, 147–48
temperature changes, 3–4, 75–77, 79, 81–82, 85
temporal risks, 102
temporal/spatial proximities, xvii, 93*fig.*, 95–96
Teng, F., 140
Texas, 5
"The Historical Roots of Our Ecological Crisis" (White), 168

Thibodeau, P.H., 61–62
Thinking Like a Mountain (Leopold), 108
Thriving beyond Sustainability: Pathways to a Resilient Society (Edwards), 163–64
throwaway societies, 57
tipping points, 4, 27, 94, 149
Titanic (ship), 2, 27, 98, 150
Titchener circles, xxi, xxiii, xxiv
tokenism, 102*fig.*, 104
tornadoes, 3, 96
trade, 21, 148
tragedies, xxiii, 1–9, 27–28, 29, 34, 63–64, 94, 150
transitions: everyday transitions, 164–65; Great Transitions, 106–7, 108–9, 148–50; to land ethic worldview, 26, 53, 151–52, 163–64, 167, 169–70, 172; the Oberlin Project and, 163; Schumacher College and, 162–63; worldviews and, 65–66
Transition Towns, 163
Transition Town Totnes, 160–64
transportation, 24, 57, 101, 102, 127, 153, 155, 156–57, 160, 168
Triandis, H.C., 17–18
tropical diseases, 8
Turkmenistan, 6
Tuvalu, 7
Twenge, Jean, 93, 99

uncertainties, 70–71, 74, 81, 82*fig.*, 88–89
United Church of Christ, 169
United States: carbon (CO2) emissions and the, 57; Central Park and the, 167; cleverness v wisdom and the, 100; climate change and the, 8, 90; cultural lenses and the, 40; distance and the, 49, 170; droughts and the, 5–6, 97; economics and the, 103; ego orientations and the, 49; green spaces and the, 167; happiness/income/wealth and the, 56, 58–59; individualistic worldviews and the, 52; mastery-oriented individualistic worldviews and the, 45, 48–50, 51*fig.*, 72–73, 93; optimism bias and the, 82–83; perceived inequities and the, 103; right-wing political ideologies and the, 85–86; Schwartz, Shalom and the, 48–52; self-enhancement/self-esteem/separateness and the, 49; superiority and the, 49–50; Transition Towns and the, 163; urban lifestyles and the, 42. *See also individual cities; individual states*
United States worldview, 32–64; African Americans and the, 32; analytic thinking and the, 40, 61–62; anticipation and the, 21; anti-intellectualism and the, 84–85; barriers and the, 73*fig.*; clarity and the, 106; climate change and the, xx, xxiii, 12, 21, 25, 27, 29, 33, 64–66; consumerism and the, 56, 129; creativity and the, 66; cultural scripts and the, 37–40; distance and the, 34–37, 40–52, 127; distortions and the, xx, 12, 21, 25, 29, 33, 64–65, 66, 106; eco orientations and the, 52–55; ego orientations and the, 34*fig.*, 52–55, 65, 72–73, 138; environmental systems and the, xix; environmental threats and the, 33; Fromm, E. and the, 22; Great Transitions and the, 108; happiness and the, 21, 22, 25, 33, 55–61; hierarchy and the, 40, 44–52, 108; holistic thinking and the, 40, 61–62; human/nature split and the, 64–65; impropriety denials and the, 62–64; indifference and the, 34–37; individualistic worldviews and the, xxiii, 40–44; just-world beliefs and the, 40, 62–64; lifestyles and the, 33; materialism and the, 33–34, 55–61, 108, 129–30; misperceptions and the, 12; mitigation and the, 11, 73*fig.*; motivations and the, 11, 66; Muller-Lyer illusion and the, xxii; natural world and the, 33–34, 36–37, 64–65, 127; obstacles and the, 11, 106–7; optimism bias and the, 82–83; proenvironmental behaviors and the,

26; progress and the, 40, 55–61; psychology and the, 11; science and the, 23; self-enhancement and the, 33–34, 37–40, 65, 72–73; separateness and the, 34–37, 66, 108; sight and the, xix; societal collapse and the, 21, 33; superiority and the, 37–40, 44–52; survival and the, 33; tragedies and the, 27–28; well-being and the, 21, 22, 25, 33, 59*fig.*; "we" sense and, 40–42, 45*fig.* *See also* mastery-oriented individualistic worldviews
unjustified clarity distortions, 17–20, 29–32
urban environments, 143
urban lifestyles: climate change and, 99; distance and, 36, 52; environmental threats and, 99; event notice (stage 1) and, 75*fig.*, 77–78; individualistic worldviews and, 44, 77–78; industrialization and, 42, 77; Kaplan, Steve and Rachel and, 99; Land Ethic Scale (Mayer and Frantz) and, 114; mechanical distance and, 42; natural world and, 36, 42, 52, 77–78; place attachments and, 94; problem-solutions and, 99; United States and, 42
urban planning, xvii, 167–68
U.S. Department of Energy, 153
Uzbekistan, 6

validity, 114, 124–25
values, 17–19, 45–47, 90, 102*fig.*, 103–4, 108, 112, 126, 144, 170. *See also* Schwartz, Shalom
Vaux, Calvert, 167
victim blame, 63–64
Vietnam, 6
vision distortions, 28, 29. *See also* clarity; distortions
Vitamin N: The Essential Guide to a Nature Rich Life (Louv), 166–67
Vlek, C., 127

Walker, I. A., 127
water, 5–6, 8, 20–21
wealth, 55–61; carbon (CO2) emissions and, 55; communities and, 59; cultural worldviews and, 65; cultures and, 51*fig.*, 58; happiness and, 58–59, 65; ignorance and, 81; mastery-oriented individualistic worldviews and, 106; natural world and, 58–60, 65–66, 72, 81, 129–30; nature connectedness and, 129–30; race and, 58–59; right-wing political ideologies and, 87; Schwartz, Shalom and, 47–48, 51*fig.*; self-enhancement and, 55; social justice and, 53; Transition Town Totnes and, 161; United States and, 58–59; well-being and, 58–60, 138. *See also* consumerism; materialism
weather events, 3, 4–5. *See also individual weather events*
Weber, Elke, 10
WeCar by Enterprise, 157
well-being, 138–43; alternative worldviews and, 21, 22; belonging and, 138–39, 140, 144; communities and, 106, 138–39, 140, 160–61; cultures and, 17, 19; education and, 139; Great Transitions and the, 108; income and, 59*fig.*; individualistic worldviews and, 44; just-world beliefs and, 64; Land Ethic Scale (Mayer and Frantz) and, 139–41, 144–45; land ethic worldview and, 23, 26, 118, 138–43, 144–45, 159; marriage and, 139; mastery-oriented individualistic worldviews and, 118, 120, 138; materialism and, 58–61; moral development and, 55; natural world and, 44, 65, 72, 118–19, 120, 122–23, 132, 140; nature connectedness and, 139, 140; scales and, 122–23, 124–25, 139–41, 144–45; self-enhancement and, 138; spirituality and, 144–45; United States worldview and, 21, 22, 25, 33, 59*fig.*; wealth and, 58–60, 138. *See also* happiness
"we" sense, 40–42, 45*fig.*, 54*fig.*, 68–69, 72, 79, 92–97, 99, 110–11, 130

White, Lynn, 168, 169
Why Nations Go to War (Stoessinger), 83
wildfires, 3, 5, 97
wildlife conservation groups, 133*fig.*, 136. *See also* environmental organizations
Wilson, E. O., 116
wisdom v cleverness, xvi, 100
wolves, 108, 109, 111, 112–13, 131
World Commission on Environment and Development (WCED), 26
World Health Organization (WHO), 140
worldview critiques, xxiv, 21–24
worldviews: actions and, xix, 24–25, 65, 106; African Americans and, 32; behavior changes and, 24; clarity and, 29; climate change and, xix, 12, 27; cultural schemas, 13*fig.*, 17–21, 30; cultural scripts and, 39; cultures and, 32, 43; distortions and, 16–17, 19–20, 29, 32, 64–65, 106; Easter Island and, 20; ethnocentric beliefs, 93*fig.*, 96; generations and, 43; harmony and, xxiii, 28, 65–66; helping and, 72; Hierarchy/Mastery (Schwartz model) and, 47; just-world beliefs, 40, 62–64, 93*fig.*, 96, 106; natural world and, 12, 24–25, 28, 37, 72, 106–7, 114; nature connectedness and, 107; New Environmental Paradigm and, 23; policies and, xix, 24; psychology as a foundational science and, 21–24, 28; science and, xxiv, 21–22, 23, 109; the Seekers (cult group) and, 31–32; self-absorption and, 66; separateness and, 106; systems thinking and, 66; transitions and, 65–66. *See also* cultural lenses; land ethic worldview; mastery-oriented individualistic worldviews; United States worldview
World War I, 83
World War II, 35, 36, 37–38
Wright, Ronald, 27
Wyoming, 5

Zelenski, J. M., 120–21, 133–34, 135, 136, 137, 141, 142, 143